Recent Advances in Renewable Energy

(Volume 2)

(Application of Flexible AC Transmission System Devices in Wind Energy Conversion Systems)

Authored by

Ahmed Abu-Siada, Mohammad A. S. Masoum & Yasser Alharbi
Electrical and Computer Engineering Department, Curtin University, Australia

Farhad Shahnia
School of Engineering and Information Technology, Murdoch University, Australia

A.M. Shiddiq Yunus
Energy Conversion Study Program, Mechanical Engineering Department, Center of Sustainable Energy and Smart Grid Applications, State Polytechnic of Ujung Pandang, Indonesia

General:

1. Any dispute or claim arising out of or in connection with this License Agreement or the Work (including non-contractual disputes or claims) will be governed by and construed in accordance with the laws of the U.A.E. as applied in the Emirate of Dubai. Each party agrees that the courts of the Emirate of Dubai shall have exclusive jurisdiction to settle any dispute or claim arising out of or in connection with this License Agreement or the Work (including non-contractual disputes or claims).

2. Your rights under this License Agreement will automatically terminate without notice and without the need for a court order if at any point you breach any terms of this License Agreement. In no event will any delay or failure by Bentham Science Publishers in enforcing your compliance with this License Agreement constitute a waiver of any of its rights.

3. You acknowledge that you have read this License Agreement, and agree to be bound by its terms and conditions. To the extent that any other terms and conditions presented on any website of Bentham Science Publishers conflict with, or are inconsistent with, the terms and conditions set out in this License Agreement, you acknowledge that the terms and conditions set out in this License Agreement shall prevail.

Bentham Science Publishers Ltd.
Executive Suite Y - 2
PO Box 7917, Saif Zone
Sharjah, U.A.E.
Email: subscriptions@benthamscience.org

BENTHAM SCIENCE

CONTENTS

PREFACE

Due to the continuous resources' reduction and cost increase of conventional fossil fuel along with the global trend to decrease the greenhouse effect, clean energy production from renewable sources has been given a global great concern. Among renewable energy sources, wind energy conversion systems have received a worldwide notable attention. It is expected that more than 10% of the global electricity demand will to be generated by wind energy conversion systems by the year 2020. During their early implementation stage, wind turbines were to be disconnected during abnormal and fault conditions within the electricity grid it is connected to. Owing to the fact that current wind installations supply a significant portion of the load demand, disconnecting windfarms may lead to business interruption and discontinuity of power supply to the end user. As such, transmission line operators have developed strict grid codes that wind turbine generator must meet to maintain its connection to support the grid during various fault conditions. To comply with these codes, flexible AC transmission systems have been widely used with current wind energy conversion systems to modulate reactive and/or active power at the point of common coupling of the wind turbine generator and the grid.

This book presents the applications of various flexible ac transmission system devices to wind energy conversion systems. Devices such as unified power flow controllers, superconducting magnetic energy storage and static synchronous compensator are covered in this book. Topologies, control systems along with case studies of the aforementioned devices are presented and discussed.

This book will be useful for postgraduate research students, upper-division electrical engineering students and practicing engineers.

Ahmed Abu-Siada
Electrical and Computer Engineering Department
Curtin University
Kent St, Bentley WA 6102
Australia

CONFLICT OF INTEREST

The authors declare no conflict of interest, financial or otherwise.

ACKNOWLEDGEMENTS

Declared none.

CHAPTER 1

Overview of Wind Energy Conversion Systems and Flexible AC Transmission Systems

Abstract: Flexible AC transmission system (FACTS) is a technology that consists of a variety of power electronic devices which was developed with the aim of controlling both power and voltage at certain locations of the electricity grids during disturbances, improving the existing transmission line capacity and providing a controllable power flow for a selected transmission direction. This chapter provides a general overview for variable FACTS devices, concepts and topologies. It also provides brief information about various wind energy conversion systems.

Keywords: Flexible AC transmission systems, Wind energy conversion systems.

INTRODUCTION

The increase in human population in the last few decades has been associated with concerns as to the corresponding rise in demand for life-supporting resources such as water, food and electrical power [1]. As for electrical power, the unparalleled industrial and technology advances are other factors that call for increasing demand in electrical consumption [2]. Globally, the power generation sector is facing significant challenges to meet the increasing demand for power. To date, conventional energy sources including oil, gas and coal are the world's main sources of energy. Unfortunately, these fossil fuel resources are associated with emissions that can severely harm the environment, with the symptoms being as air pollution, climate change, oil spills and acid rain [3]. Interest in harnessing the benefits of renewable energy has been increasing steadily due to its advantages, which include sustainability, environmental friendly nature and affordable cost. Solar, geothermal and wind resources are among the most promising renewable energy alternatives [4, 5].

As a natural energy source, the radiation delivered by the sun is a promising renewable energy source of electrical power [6]. Nowadays, solar energy is one of the favourable energy alternatives. Photovoltaic (PV) technology is used for converting solar energy to electrical energy [7]. Globally, PV installation has contributed to about 177 GW of electrical power in 2014, and it is expected to deliver more than 1% of the total global electricity demand by the end of 2017

Ahmed Abu-Siada, Mohammad A.S. Masoum, Yasser Alharbi, Farhad Shahnia & A.M. Shiddiq Yunus

[8]. The geothermal power has the advantage of using fewer infrastructure elements for electrical power generation when compared with other energy sources such as coal or nuclear power [9]. In 2015, the world's installed capacity of geothermal power reached 12.635 MW, and this is expected to reach about 21.441 MW by 2020 [10]. Since the early stages of using natural energy resources, wind power has been considered as a main renewable energy source. And nowadays, there is a rapid increase in the utilization of wind energy [11], which has led to significant advancement in wind energy technology, including wind turbine design and sizing, and integration of wind turbines with existing electricity grids.

WIND ENERGY SYSTEM

Wind energy has become one of the most popular renewable energy sources worldwide. In 2014, there was an additional of 51,473 MW of new wind power capacity that was brought into service [12]. Fig. (**1.1**) illustrates the top 10 installed wind power capacity worldwide during the period from January to December 2014. The diagram shows that China has the highest installed wind power capacity with 23,196 MW generation, followed by Germany at 5,279 MW and USA at 4,279 MW [12].

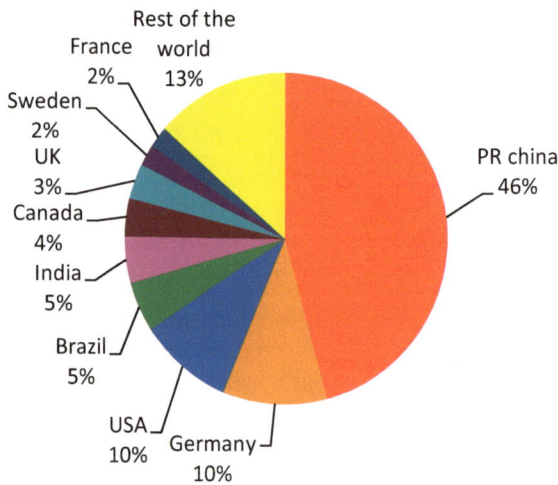

Fig. (1.1). Distribution of the top 10 installed wind capacity in 2014.

Figure (**1.2**) shows the magnitude of the globally installed wind- capacity between 1996 and 2014. It can be seen from Fig. (**2.2**) that the capacity increased from 197,943 MW 369,597 MW over this period.

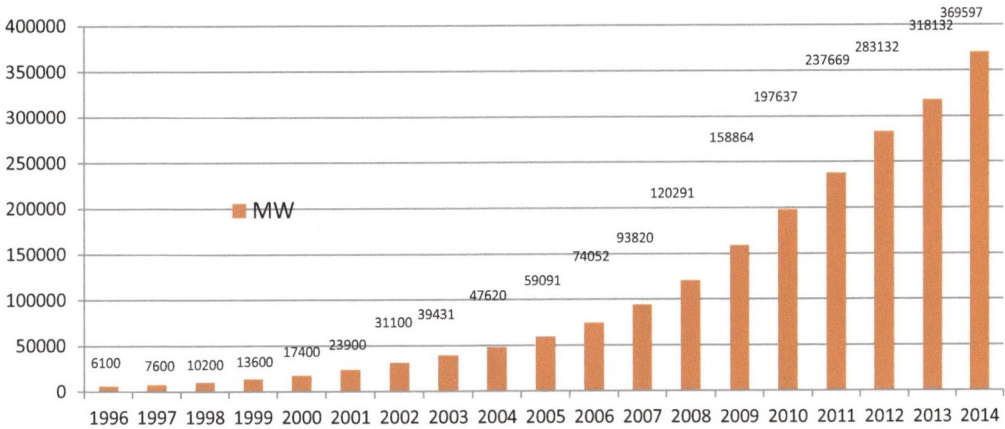

Fig. (1.2). Global production of wind power between 1996 and 2014.

The installed wind power capacity in Australia reached 3,806 MW by the end of 2014, and the installed wind power capacity in the year 2014 was 13% less than that installed in 2013 [12]. Despite the decrease in the newly installed capacity in 2014, the wind energy market is leading Australia towards its goal of using renewable energy to supply 20% of the power requirements by the year 2020 [12].

WIND TURBINE

Wind turbines capacity ranges from a few kilowatts for standalone units for houses to several megawatts in a wind farm. Small wind turbines are usually rated below 300kW and have the capability to be combined with other energy sources as generation system at farms and houses to support the need for electrical power. However, the integration of small wind turbine with existing grids is difficult and costly [13]. The wind turbine size and rating have been increasing gradually since 1980 as shown in Fig. (**1.3**); increasing the size of the wind turbine rotor increases the amount of energy harvested by the wind turbine. In the early stage of wind turbine manufacturing, wind turbine power rating started with 50 kW and a size of 15 m rotor radius but nowadays wind turbines are designed to produce up to 7.5 MW with up to 126 m rotor diameter. A higher rating of 10 MW and associated 160 m rotor diameter is available nowadays as well [13].

WIND ENERGY CONVERSION SYSTEMS

Both fixed speed and variable speed generator can be used in wind energy conversion systems (WECS). In the early stages of the design of (WECS), wind turbines used to function at fixed speeds. Nowadays, with the new concept of generators and power electronics, variable speed wind turbines dominate the wind turbine market. Fig. (**1.4**) illustrates different configurations of wind energy conversion systems. A review of the two types of fixed and variable speed systems is given below.

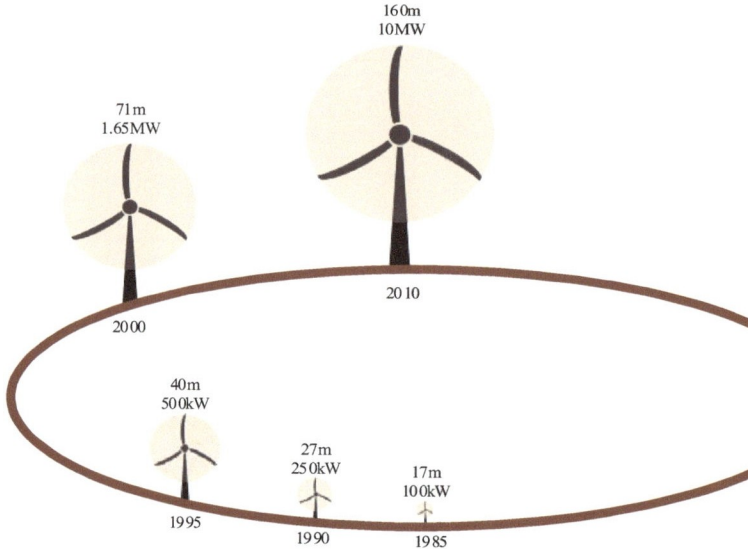

Fig. (1.3). Evolution of wind turbine size.

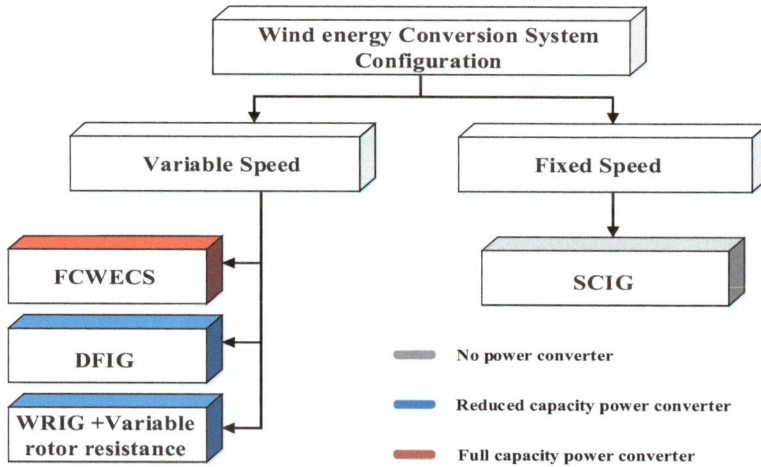

Fig. (1.4). Different Configurations of WECS.

Fixed Speed Wind Turbine

A fixed speed wind turbine (Fig. **1.5**) comprises a generator that is directly coupled to the power network and connected to the wind turbine through low-speed shaft, gearbox and high-speed shaft [14].

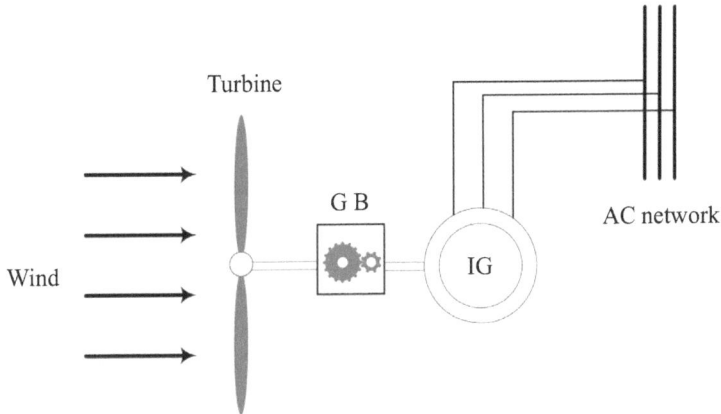

Fig. (1.5). Fixed speed WECS configuration.

Fixed speed WECS has the advantages of being simple, inexpensive in addition to the fact that it does not call for a power electronic interface. However, fixed speed wind turbines suffer from the limitation of controlling the power quality, uncontrollable reactive power compensation and high mechanical stress on the shaft sections [14]. The oldest wind energy conversion system topology employed a fixed speed generator (*e.g.* synchronous generator) that is coupled directly to the AC network, including mechanical dampers in the drive train. Modern fixed speed wind energy conversion systems use induction generators [15]. Fig. (**1.6**) shows a conceptual scheme of the first fixed speed wind energy conversion system.

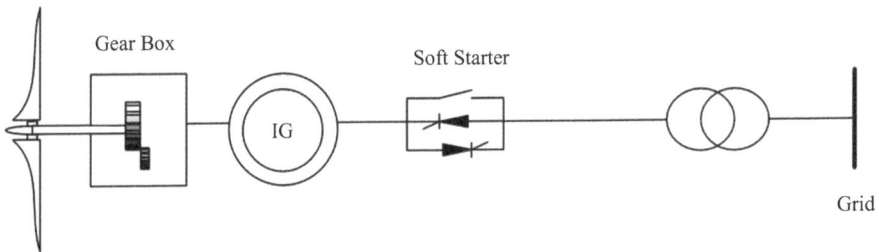

Fig. (1.6). Typical configuration of the first generation of WECS.

Variable Speed Wind Turbine

Technology advancement has driven the wind turbine operation from being of fixed speed mode to variable speed. A variable speed wind turbine consists of a generator driven by a power converter, which facilitates the variable speed operation mode and aids in improving the WECS dynamic performance [14]. Owing to its variable speed mode, more power can be captured, enhanced power quality can be achieved and reduced mechanical stress on the drive train can be accomplished by wind power generators [14 - 16]. Comparison between the fixed speed wind turbine and variable speed wind turbine is summarized in Table **1.1**.

Table 1.1. Comparison of fixed and variable speed-based WECS.

	Variable Speed Wind Turbine	**Fixed Speed Wind Turbine**
Advantages	-Maximized captured power -Simple pitch control -Reduced mechanical stress -Provides dynamic compensation for torque and power pulsation -Improved power quality -Reduced acoustic noise	-Low cost -Simple structure -Low maintenance
Disadvantages	-High cost -Complex control system	-High mechanical stress -High power fluctuations to the grid -Relatively low energy conversion

Partly Variable Speed Wind Turbine

Partly variable speed wind turbines (Fig. **1.7**) or the so-called type B wind turbines operate in a limited variable speed mode. In this concept, the generator is directly coupled to the AC network and the generator rotor is connected to a variable resistance to control the generator speed. Depending on the variable resistance size, the slip can be increased by up to 10% that allows operation at a partly variable speed in the super synchronous range (*i.e.* up to 10% above the rated speed). The Danish manufacturer, Vestas, used this design feature in limited variable speed wind turbines since the mid1990's [14, 15].

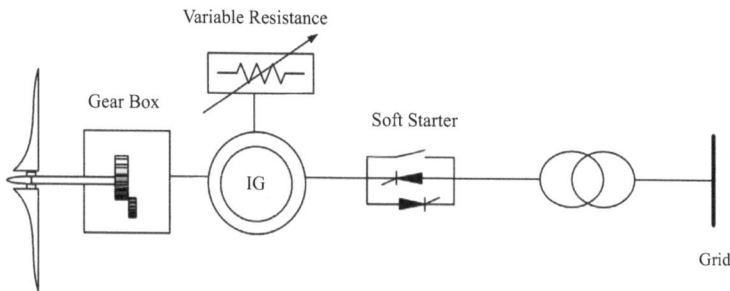

Fig. **(1.7)**. Typical configuration of partly variable speed WECS.

Full Converter Variable Speed Wind Turbine

A full converter variable speed wind turbine that based on a multi-pole synchronous generator is shown in Fig. (**1.8**). In this type, the generator is coupled to the AC network through a full-scale converter station that facilitates the variable speed operation of the wind turbine. The converter station is a combination of grid side converter and generator side converter connected back to back via a dc link. The generator's electrical frequency changes with the change in the wind speed, whereas the power network frequency remains unaffected [14].

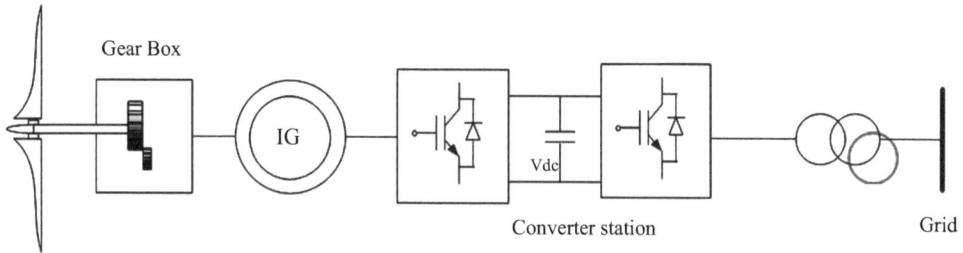

Fig. (1.8). Typical configuration of full converter variable speed WECS.

Doubly Fed Induction Generator Wind Turbine

Typical configuration of a doubly fed induction generator (DFIG) wind turbine is shown in Fig. (**1.9**). Among the variable speed wind turbine generators, DFIG is the most popular technology currently dominating the market of wind turbines, because of its superior advantages over other wind turbine technologies [17]. In this concept, the stator circuit is coupled directly with the power network through a coupling transformer while a back-to-back partial-scale voltage source converter (VSC) connects to the rotor circuit to the grid via the coupling transformer. The VSC facilitates a decoupled control of the generator's active and reactive power [15].

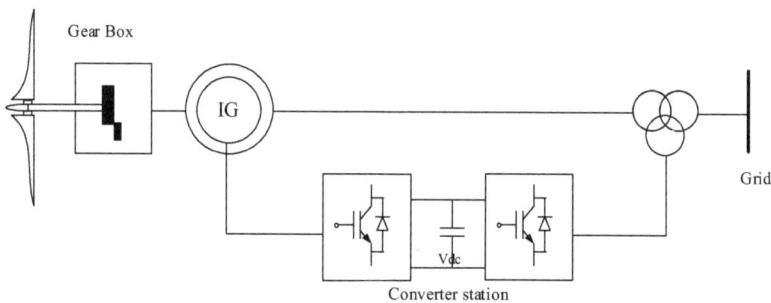

Fig. (1.9). Typical configuration of Doubly Fed Induction Generators WECS.

In this configuration, the power can be supplied to the grid through both the stator and the rotor, while the rotor can also absorb power in some operational modes that are determined by the generator speed. The produced power is supplied to the existing grid from the rotor via the VSC during the super-synchronous operational mode. On the other hand, the rotor absorbs power from the network through the converter station during the sub-synchronous operational mode as can be seen in Fig. (**1.10**).

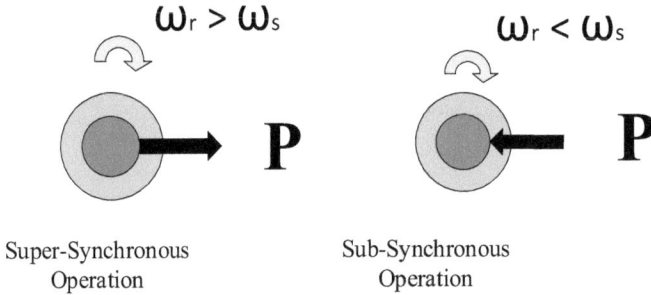

Fig. (1.10). Super-synchronous and sub-synchronous operational modes.

Impact of Wind Energy Integration into Electricity Grid

The integration of wind energy into electricity grid has been given more concern in recent years due to the significant increase in wind power generation level [18]. The high penetration of WECS into electricity grid leads to technical and power quality issues that need more consideration by the system operators [18]. Some of these issues include:

Power Variation

Significant change in the power may occur because of the changing nature of the wind gust or the WECS tower shadow effects. Switching operation can also lead to power variation phenomena [19].

Harmonics

The use of power electronic converter stations to interface wind turbine generator with the grid results in distorted voltage and current waveforms [20].

Voltage Fluctuation

Voltage variation at the point of common coupling (PCC) connecting wind turbine generator with the grid may take place due to the intermittent nature of wind and if an adequate control scheme to regulate the voltage level is not adopted. Other causes for voltage variation include significant load changing and fault conditions [21].

Flickers

Flickers may arise because of a rapid and continuous change in the wind generated power [20].

Fault Ride through and Grid Codes

Due to the significant increase in wind turbine generators (WTGs) and the global trend to establish reliable smart grids, the transmission system operators (TSOs) require connection of WTGs with existing grids to provide power support during maintenance and fault circumstances of the latter. Accordingly, grid codes have been established in many countries to comply with these new requirements. There are many international codes related to the fault ride through (FRT) capability of WTGs [22 - 26]. The low voltage ride through (LVRT) grid codes for UK, Spain, Denmark, Germany, Australia, China and the US are shown in Fig. (**1.11**) as examples.

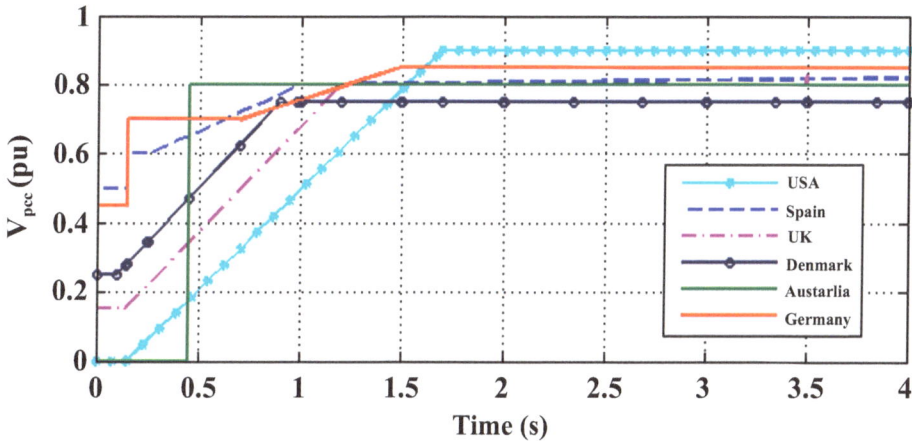

Fig. (1.11). Nominated countries low voltage ride-through grid (LVRT) codes.

The allowed voltage sag at the PCC of the United States grid code is 0 pu, which lasts for a duration of 0.15s from the occurrence of the fault after which the LVRT profile increases linearly during the following 1.5s to 0.9 pu at which the voltage level is maintained [22].

The minimum acceptable voltages sag at the PCC for the Spanish grid code at the instant of fault occurrence is 0.5 pu, which lasts for 0.15s after which it increases to 0.6 pu and lasts for 0.1s. The LVRT profile then ramps to 0.8 pu during the next 0.75s and remains at this level for 3s [22]. The Australian grid code requires WTG to withstand a PCC voltage level of 0 pu for 3s then LVRT increases to 0.8

pu [26]. WTGs are to be disconnected from the grid in case of voltage levels at the PCC fall outside the area bounded by the LVRT margins of the grid codes.

Fig. (**1.12**) shows the technical requirements for the high voltage ride through (HVRT) capability of wind turbine generators in some countries. The allowed voltage swell at the PCC of the US grid code is 1.2 pu, which lasts for a duration of 1s from the fault occurrence. After that, the HVRT profile decreases by 0.05 pu every 1s during the following 3s, after which the voltage at the PCC has to be maintained within a safety margin of 0.05 pu from the nominal value [22].

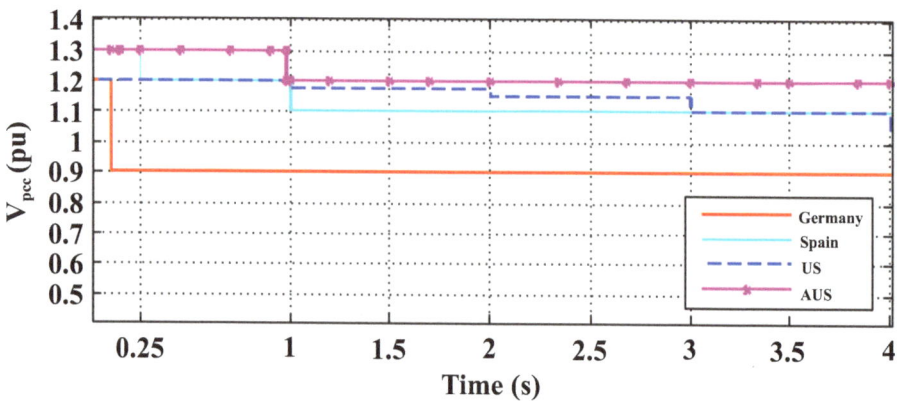

Fig. (1.12). Nominated countries High voltage ride-through grid codes.

Table 1.2. Some parameters for Grid codes in some countries [23, 25, 26].

| Country | LVRT | | | | HVRT | |
| | During Fault | | Fault Clearance | | During Swell | |
	V_{min} (pu)	T_{max} (s)	V_{min} (pu)	T_{max} (s)	V_{max} (pu)	T_{max} (s)
Denmark	0.25	0.1	0.75	0.5	NA	NA
Sweden	0.25	0.25	0.9	0.25	NA	NA
Germany	0	0.15	0.9	1.5	1.2	0.1
Spain	0	0.15	0.85	1	0.3	0.25
USA	0	0.15	0.9	1	1.2	1
Australia	0	0.45	0.8	0.45	1.3	0.98
China	0.2	0.625	0.9	3	NA	NA

The maximum voltage rise at the PCC for the Spanish grid code at the instant of fault occurrence is 1.3 pu which remains for 0.25s after which it decreases by 0.1

pu that lasts for 1s. Then the voltage level at the PCC has to be maintained within a safety margin of 0.1 pu above the nominal value [22]. The Australian standard allows the voltage to increase by 1.3 pu for 0.98s after that the voltage has to be limited to 1.2 pu [26]. Table **1.2** lists the LVRT and HVRT codes in 7 countries.

OVERVIEW OF FACTS DEVICES

The concept of FACTS (flexible AC transmission systems) was envisioned in the late of 1980s [27]. The technology consists of a variety of power electronic devices with the aim of controlling both power and voltage at a certain location of the electricity grids during disturbances. In general, the FACTS devices were invented to improve the existing transmission line capacity and provide a controllable power flow for a selected transmission direction [28].

The FACTS devices are primarily divided into two groups. The first group involves quadrature tap-changing transformers, including conventional thyristor-switched capacitors and reactors, and the second group involves voltage source converters based on gate turn-off (GTO) thyristor-switched converters [29]. The first group has introduced Thyristor-Controlled Phase Shifter (TCPS), Static Var Compensator (SVC) [30] and the Thyristor- Controlled Series Capacitor (TCSC) [31]. The second group has resulted in the Static Synchronous Series Compensator (SSSC) [31], the Static Synchronous Compensator (STATCOM) [32], the Interline Power Flow Controller (IPFC) [33] and the Unified Power Flow Controller (UPFC) [33]. Each generation of the FACTS devices has its own performance and characteristics [34].

Capacitor and reactor banks along with fast solid-state switches are used in the first group of FACTS devices, which can be connected in series or shunt with the power system to compensate for the reactive power at the PCC. However, real power exchange with the system is not possible with such systems.

The devices of the FACTS group that is based on voltage source converter (VSC) utilises self-commutated converters equipped with GTO thyristor switches. Through a proper control scheme, this group is able to generate capacitive and inductive reactance internally [29]. In addition to the independent reactive power control, the voltage source converter can be integrated with energy storage system to enable decoupled control of active and reactive power exchange with the system it is connected to [28]. Fig. (**1.13**) illustrates an overview of the main FACTS devices.

To explain the reactive power compensation process of a shunt FACTS, a shunt connected compensator (simulated as an ideal AC current source) is connected to the middle of a lossless transmission line for the system shown in Fig. (**1.14**).

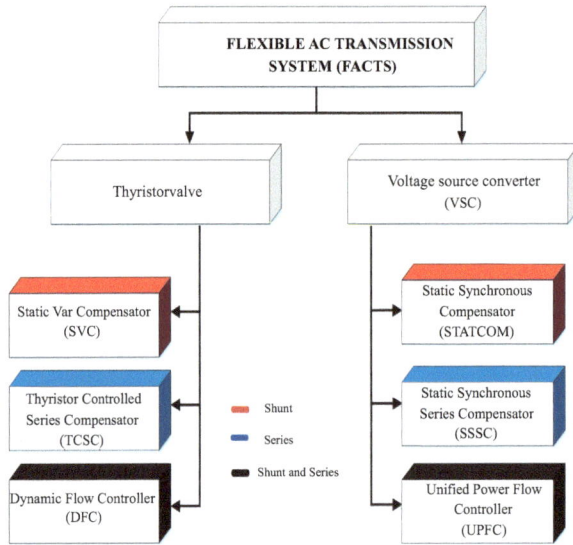

Fig. (1.13). Overview of the main FACTS Devices.

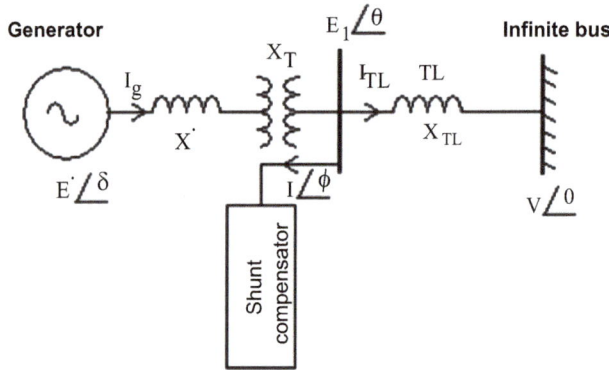

Fig. (1.14). One machine – infinite bus system with shunt compensator.

For an assumed lossless system shown in Fig. (**1.14**), applying Kirchhoff's current law at the compensator bus gives:

$$\frac{E\angle\delta - E_1\angle\theta}{j(X_g+X_T)} = \frac{E_1\angle\theta - V\angle 0}{jX_{TL}} + I\angle\phi \qquad (1.1)$$

From (**1.1**), the voltage at the compensator bus $E_1\angle\theta$ can be written as:

$$E_1\angle\theta = \frac{1}{X}[X_{TL}E\angle\delta + (X_g + X_T)V + X_{TL}(X_g + X_T)I\angle\phi \qquad (1.2)$$

where $X = X_g + X_T + X_{TL}$

The apparent power of the generator is given by:

$$P_g + jQ_g = \frac{EE_1}{X_g + X_T}\sin(\delta - \theta) + j\frac{E^2 - EE_1\cos(\delta - \theta)}{X_g + X_T} \tag{1.3}$$

Eliminating E_1 and θ from (**1.3**) using (**1.2**) gives:

$$P_g = \frac{EV}{X}\sin\delta + \frac{X_{TL}E}{X}I\sin(\delta - \phi) \tag{1.4}$$

$$Q_g = \frac{E^2 - EV\cos\delta}{X} - \frac{X_{TL}E}{X}I\cos(\delta - \phi) \tag{1.5}$$

The first terms in (**4**) and (**5**) give the active and reactive power transfer through the transmission line without the shunt compensation, respectively. When the shunt compensator is connected, additional power represented by the second term in both equations can be provided. By controlling the current magnitude (I) and/or the phase angle (φ) at the compensator terminals, the power flow from the generator to the infinite bus can be modulated. It can be observed that when δ-φ = 0, the shunt compensator delivers or absorbs reactive power depending on the direction of the current. On the other hand when δ-φ = π/2, the compensator only modulates the active power at the point of common coupling. For any other value, the compensator can modulate both active and reactive power.

FACTS devices introduce a wide operational range to increase the thermal limits of a power system and improve its stability. FACTS devices have been used for:

- Active and reactive Power control [35]
- Voltage control stability and Sub synchronous resonance mitigation [36]
- Power oscillation damping [37]
- Transients and dynamics stability [38]
- Fault current limiting [39]
- Flicker mitigation [40]
- Power system security enhancement [41]

Investments into such complex devices have to carefully consider many aspects such as the practical requirements and benefits of the application. Some FACTS devices can perform multi tasks as given in Table **1.3** below.

Among the FACTS devices listed in Table **1.3**, UPFC has the capability to cover all the listed applications due to its ability to control all the parameters affecting the transmission line power flow, including voltage, phase angle and impedance [33].

Table 1.3. FACTS Applications and cost Comparison

	SVC	STATCOM	TSCS	SSSC	UPFC
Reactive power control	x	x	x	x	•
Active power control	x	x	x	x	•
Voltage control	•	•	x	x	•
Voltage stability improvement	•	•	x	•	•
Power oscillations damping	•	•	•	•	•
Transients and dynamic stability	•	x	•	•	•
Fault current limiter	x	x	•	•	•
Var compensation	•	•	x	x	•
Approximate Costs (US $) [42]	40/kVar	50/kVar	40/kVar	20/kVar	50/kVar

SUMMARY

This chapter reviews some of the existing renewable energy sources, with focus on wind energy conversion systems. In addition, fault-ride through capability of the wind turbine generator and some grid codes are discussed. The basic concepts of various flexible AC transmission system (FACTS) devices were introduced in this chapter along with their applications in power systems. A comparison between the FACTS devices based on their applications and costs was also presented.

REFERENCES

[1]   H. Polinder, D-J. Bang, H. Li, and Z. Cheng, *Concept Report on Generator Topologies, Mechanical and Electromagnetic Optimization.* Project UpWind, 2007.

[2]   L.M. Hunter, and S. Labor, *Population Program, and S. Population Matters Project, Environmental Implications of Population Dynamics.* RAND Corporation: Santa Monica, CA, USA, 2001.

[3]   I.P. Change, *Climate Change 2014: Mitigation of Climate Change.* Cambridge University Press, 2015.

[4]   M. Kaltschmitt, N.J. Themelis, L.Y. Bronicki, L. Söder, and L.A. Vega, *Renewable Energy Systems.* Springer New York, 2012.

[5]   B. Sorensen, *Renewable Energy: Physics, Engineering, Environmental Impacts, Economics & Planning.* Elsevier Science, 2010.

[6]   G. Boyle, *Renewable Energy: Power for a Sustainable Future.* OUP Oxford, 2012.

[7]   A.E. Resources, N.R. Council, and N.A. Engineering, *Electricity from Renewable Resources: Status, Prospects, and Impediments.* National Academies Press, 2010.

[8]   H.J. Boenig, and J.F. Hauer, "Commissioning Tests Of The Bonneville Power Administration 30 MJ

Superconducting Magnetic Energy Storage Unit", *IEEE Trans. Power Apparatus Syst,* vol. PAS-104, pp. 302-312, 1985.
[http://dx.doi.org/10.1109/TPAS.1985.319044]

[9]   R. DiPippo, "Geothermal Power Plants - Principles, Applications, Case Studies and Environmental Impact (2nd Edition), Elsevier",

[10]   R. Bertani, "Geothermal Power Generation in the World 2010–2014 Update Report", *Proceedings World Geothermal Congress 2015* Melbourne, Australia, 2015.

[11]   P. Musgrove, *Wind Power.* Cambridge University Press: New York, 2010.

[12]   "Global Wind Report 2014", Available: http://www.gwec.net/wp-content/ uploads/2015/03/ GWEC_Global_Wind_2014_Report_LR.pdf

[13]   B. Wu, Y. Lang, N. Zargari, and S. Kouro, "Introduction," in Power Conversion and Control of Wind Energy Systems, ed: John Wiley and Sons, Inc., 2011, pp. 1-23.

[14]   T. Ackermann, *Wind Power in Power Systems.* Wiley, 2012.
[http://dx.doi.org/10.1002/9781119941842]

[15]   V. Akhmatov, *Induction Generators for Wind Power.* Multi-Science Pub., 2005.

[16]   B. Fox, Wind Power Integration: Connection and System Operational Aspects: Institution of Engineering and Technology, 2007.
[http://dx.doi.org/10.1049/PBPO050E]

[17]   H. Polinder, J.A. Ferreira, B.B. Jensen, A.B. Abrahamsen, K. Atallah, and R.A. McMahon, "Trends in Wind Turbine Generator Systems", *Emerging and Selected Topics in Power Electronics, IEEE Journal of,* vol. 1, pp. 174-185, 2013.
[http://dx.doi.org/10.1109/JESTPE.2013.2280428]

[18]   M. Stiebler, "Grid Integration and Power Quality", In: *Wind Energy Systems for Electric Power Generation.,* S.B. Heidelberg, Ed., , 2008, pp. 147-170.
[http://dx.doi.org/10.1007/978-3-540-68765-8_7]

[19]   M. Stiebler, *Wind Energy Systems for Electric Power Generation.* Springer: Berlin, 2008.

[20]   E.F. Fuchs, and M.A. Masoum, "Power Quality in Power Systems and Electrical Machines", Elsevier, 2008.

[21]   M. Stiebler, "Wind Energy Systems", In: *Wind Energy Systems for Electric Power Generation.,* S.B. Heidelberg, Ed., , 2008, pp. 81-113.
[http://dx.doi.org/10.1007/978-3-540-68765-8_5]

[22]   Alt, x, M. n, Go, O. ksu, R. Teodorescu, *et al.,* "Overview of recent grid codes for wind power integration," in *Optimization of Electrical and Electronic Equipment (OPTIM), 2010 12th International Conference on,* 2010, pp. 1152-1160.

[23]   C.E. Institute, "Technical Rule for Connecting Wind Farm into Power Network", 2009.

[24]   M. Mohseni, and S.M. Islam, "Review of international grid codes for wind power integration: Diversity, technology and a case for global standard", *Renew. Sustain. Energy Rev.,* vol. 16, pp. 3876-3890, 2012.
[http://dx.doi.org/10.1016/j.rser.2012.03.039]

[25]   M. Tsili, and S. Papathanassiou, "A review of grid code technical requirements for wind farms", *Renewable Power Generation, IET,* vol. 3, pp. 308-332, 2009.
[http://dx.doi.org/10.1049/iet-rpg.2008.0070]

[26]   A. Western Power, *Technical Rules,* ed 2011.

[27]   L. G. c. E.-H. Narain G. Hingorani, consulting editor., Understanding FACTS : concepts and technology of flexible AC transmission systems New York: IEEE Press 2000.

[28]   P. Therond, P. Cholley, D. Daniel, E. Joncquel, L. Lafon, and C. Poumarede, "FACTS research and development program at EDF," in Flexible AC Transmission Systems (FACTS) - The Key to Increased Utilisation of Power Systems, IEE Colloquium on (Digest No.1994/005), 1994, pp. 6/1-615.

[29]   X-P. Zhang, C. Rehtanz, and B. Pal, "FACTS-Devices and Applications," in Flexible AC Transmission Systems: Modelling and Control, ed: Springer, 2012, pp. 1-30.
       [http://dx.doi.org/10.1007/978-3-642-28241-6_1]

[30]   R. Mathur, and R. Varma, "SVC Applications", in Thyristor-Based FACTS Controllers for Electrical Transmission Systems, ed: Wiley-IEEE Press, 2002, pp. 221-276.

[31]   N. Hingorani, and L. Gyugyi, "Static Series Compensators: GCSC, TSSC, TCSC and SSSC", in Understanding FACTS:Concepts and Technology of Flexible AC Transmission Systems, ed: Wiley-IEEE Press, 2000, pp. 209-265.

[32]   N. Hingorani, and L. Gyugyi, "Static Shunt Compensators: SVC and STATCOM", in Understanding FACTS:Concepts and Technology of Flexible AC Transmission Systems, ed: Wiley-IEEE Press, 2000, pp. 135-207.

[33]   N. Hingorani, and L. Gyugyi, "Combined Compensators: Unified Power Flow Controller (UPFC) and Interline Power Flow Controller (IPFC)",

[34]   N.G. Hingorani, and L. Gyugyi, *Understanding FACTS: Concepts and Technology of Flexible AC Transmission Systems.* Wiley, 2000.

[35]   D.J. Gotham, and G.T. Heydt, "Power flow control and power flow studies for systems with FACTS devices", *Power Systems, IEEE Transactions on,* vol. 13, pp. 60-65, 1998.
       [http://dx.doi.org/10.1109/59.651614]

[36]   B.K. Perkins, and M.R. Iravani, "Dynamic modeling of a TCSC with application to SSR analysis", *Power Systems, IEEE Transactions on,* vol. 12, pp. 1619-1625, 1997.
       [http://dx.doi.org/10.1109/59.627867]

[37]   E.V. Larsen, J.J. Sanchez-Gasca, and J.H. Chow, "Concepts for design of FACTS controllers to damp power swings", *Power Systems, IEEE Transactions on,* vol. 10, pp. 948-956, 1995.
       [http://dx.doi.org/10.1109/59.387938]

[38]   M. Noroozian, L. Angquist, M. Ghandhari, and G. Andersson, "Improving power system dynamics by series-connected FACTS devices", *Power Delivery, IEEE Transactions on,* vol. 12, pp. 1635-1641, 1997.
       [http://dx.doi.org/10.1109/61.634184]

[39]   K. Duangkamol, Y. Mitani, K. Tsuji, and M. Hojo, "Fault current limiting and power system stabilization by static synchronous series compensator", in Power System Technology, 2000. Proceedings. PowerCon 2000. International Conference on, 2000, pp. 1581-1586 vol.3.
       [http://dx.doi.org/10.1109/ICPST.2000.898207]

[40]   M.I. Marei, E.F. El-Saadany, and M.M. Salama, "A Flexible DG Interface Based on a New RLS Algorithm for Power Quality Improvement", *Systems Journal, IEEE,* vol. 6, pp. 68-75, 2012.
       [http://dx.doi.org/10.1109/JSYST.2011.2162930]

[41]   A. Berizzi, M. Delfanti, P. Marannino, M.S. Pasquadibisceglie, and A. Silvestri, "Enhanced Security-Constrained OPF With FACTS Devices", *IEEE Trans. Power Syst.,* vol. 20, pp. 1597-1605, 2005.
       [http://dx.doi.org/10.1109/TPWRS.2005.852125]

[42]   R.M. Mathur, and R.K. Varma, "Thyristor-Based FACTS Controllers and Electrical Transmission Systems", ed: Wiley-IEEE Press, pp. 413-461.
       [http://dx.doi.org/10.1109/9780470546680]

CHAPTER 2

Applications of Unified Power Flow Controller in Wind Energy Conversion System

Abstract: Unified power flow controller (UPFC) is one of the Flexible ac Transmission System (FACTS) devices that possess the ability of modulating both active and reactive power at the point of common coupling in four quadrant operational modes. This chapter illustrates UPFC topology, controllers with some case studies for various applications of UPFC in the DFIG-based WECS. New applications for UPFC are proposed to improve the overall performance of a DFIG-based WECS during voltage sag and voltage swell events at the grid side.

Keywords: FACTS, UPFC, Active and reactive power modulation, Power system stability.

INTRODUCTION

The unified power flow controller (UPFC) is a complex power electronic device that was developed to control and optimize the power flow in electrical power transmission systems [1]. Gyugyi introduced the UPFC in 1991 as a versatile device capable of controlling all the parameters affecting the power flow in the transmission line, including voltage, impedance and phase angle [2]. As a Flexible ac Transmission System (FACTS) device, unified power flow controller allows system to be more flexible by using high-speed response active and reactive power compensations to improve the power flow of the transmission system. Thus, installing a UPFC at critical points of the transmission system will increase both the power dispatch (up to the power rating of existing generators and transformers) and the thermal limits of line conductors, by increasing the stability margin [3]. Shunt and series converters of the UPFC can control both active and reactive powers smoothly, rapidly and independently in four quadrant operational moods [4]. As shown in Fig. (**2.1**), a UPFC is mainly a combination of a static compensator (STATCOM) and a static synchronous series compensator (SSSC) coupled through a common DC link. The application of the UPFC to power systems has been extensively considered by the power industry due to its various advantages, which include smooth control of both active and reactive power of the

Ahmed Abu-Siada, Mohammad A.S. Masoum, Yasser Alharbi, Farhad Shahnia & A.M. Shiddiq Yunus

system at the PCC and its rapid and independent performance in four quadrant operational moods [5, 6].

Bus A   **Bus B**

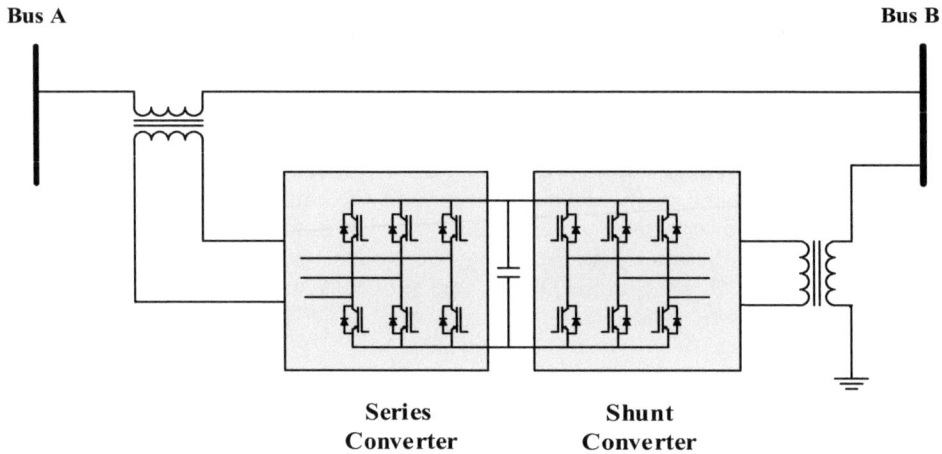

Fig. (2.1). Basic configruation of UPFC.

Shunt Converter

To compensate for the voltage of the ac network controlled bus, the shunt converter shown in Fig. (**2.2**) is used to produce a controlled imaginary current that lags or leads the fundamental component of the voltage by 90°. Inductive current (i_c) produces a positive value of the reactive power (*i.e.* $q_1 > 0$) in order to reduce the magnitude of the voltage across the ac bus, as shown in Fig. (**2.3**); this mode of operation takes place when there is a surplus reactive power generated as a result of significant load shedding that leads to voltage swell in the system. On the other hand, capacitive current (i_c) produces negative reactive power (*i.e.* $q_1 < 0$) in order to increase the amount of the voltage across the ac network bus. This operational mode takes place in case of voltage sag and short circuit faults.

Series Converter

The UPFC series converter controls the active and reactive power by controlling the voltage at the ac network bus and the current (i_s) of the transmission system. The voltage at the generator bus (v_s) is regulated by controlling the series voltage (v_c), which in turn will regulate the magnitude of the current (i_s) to meet the required active power reference P_{REF} and reactive power reference Q_{REF} (Fig. **2.4**).

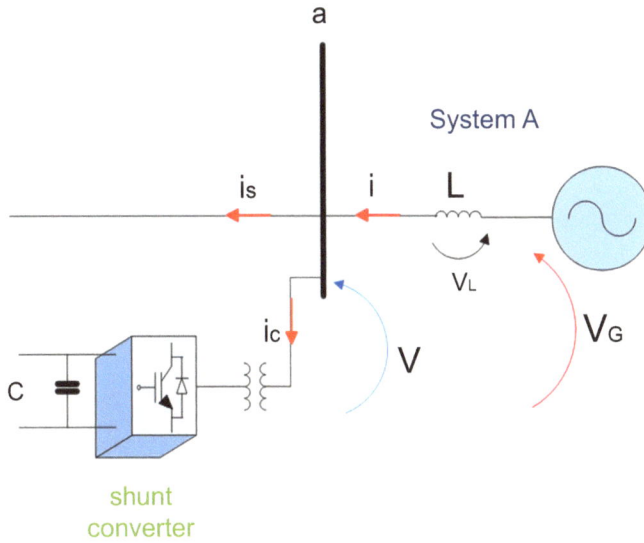

Fig. (2.2). Configurations of the UPFC shunt converter and voltage compensation technique.

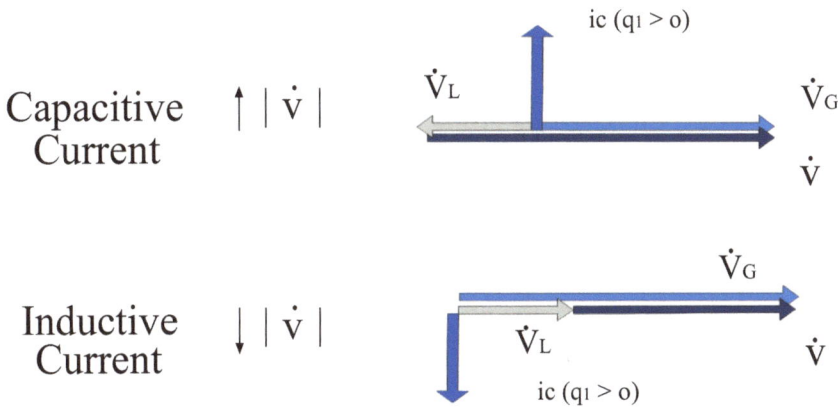

Fig. (2.3). Phasor diagram of the voltage compensation mode by the UPFC shunt converter.

To produce the required demand of the active P_{REF} and the reactive Q_{REF} power, the UPFC controls the voltage components $V_c(q_c)$ and $V_c(p_c)$ in series with the transmission line, which in turn control p and q, as shown in Fig. (**2.5**). The real power P decreases when the voltage component V_c (q_c) leads the voltage V by 90°, while the opposite occurs when the voltage V_c (q_c) lags the voltage component V by 90°. In contrast, the reactive power decreases and the current phasor I_s becomes more capacitive when the compensating voltage is greater than zero, V_c $(p_c > 0)$, whilst the reactive power increases and the current phasor I_s

becomes more inductive when the compensating voltage is less than zero, V_C ($p_C < 0$).

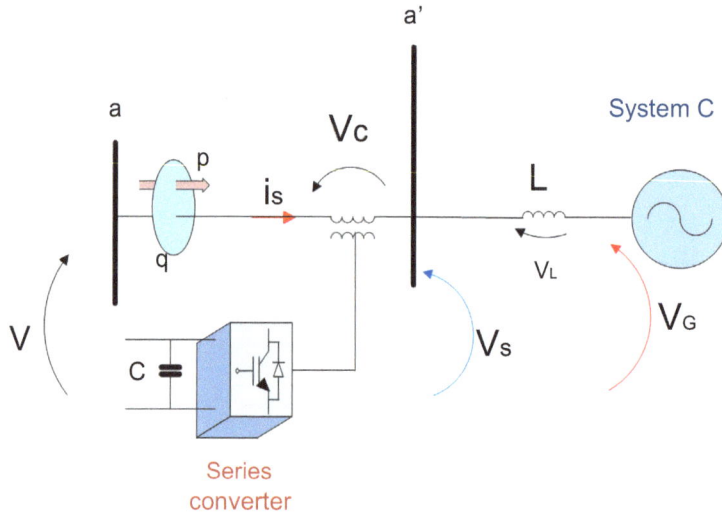

Fig. (2.4). Power flow control by the series converter of the UPFC.

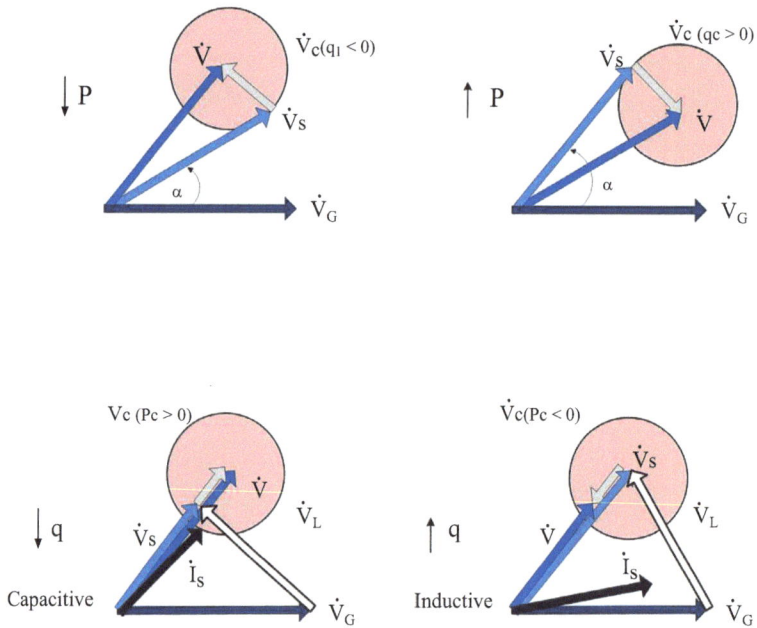

Fig. (2.5). Phasor diagram of the UPFC's series converter in the operational mode of active and reactive power control.

UPFC Applications

In an study [7], an adaptive neuro-fuzzy controller was designed for UPFC. The results show that the settling time and amplitude of the low frequency oscillations are reduced by the proposed adaptive neuro-fuzzy controller. A Lyapunov-based adaptive neural controller was proposed in the study [8] for the UPFC to improve the transient stability of the studied system. In this study, the dynamic characteristics of the UPFC were investigated and a supplementary control system was developed to suppress power swings. Damping of the power system oscillation using UPFC based on a simple PI controller was presented in the study [9]. A fuzzy-logic control scheme was presented in the study [10] to improve the power system stability during transient conditions. In an study [11], application of UPFC based on nonlinear PID and tracker controller was introduced. Table **2.1** summarizes the various applications of the UPFC along with proposed controller.

Table 2.1. Summary of the applications of UPFC along with the proposed controllers.

	Controller	Application	Ref.
1	Neuro Fuzzy	Damping Low Frequency Oscillations	[7]
2	Gravitational Search Algorithm (GSA)		[12]
3	Lyapunov	Transient Stability Enhancement	[8]
4	Neuro Controller	Damping Power System Oscillation	[13]
5			[9]
6	PI controller	Power-Flow Control Performance Analysis of UPFC	[14]
7		Real and Reactive Power Coordination for a UPFC	[15]
8	Radial Basis Function Neural Network (RBFNN)	Improving Transient Stability Performance of Power System	[16]
9	Nonlinear Variable-gain Fuzzy Controller		[10]
10	Nonlinear PID (NLPID) Controller		[11]
11	Feed-Back Linearization Controller (FBLC)		[17]
12	Nonlinear Optimal Predictive Controller		[18]
13	Based on H_2 method	Damping of Oscillation in Distribution System	[19]

Application of UPFC to Improve Overall Performance of DFIG-based WECS During Grid Side Faults

In a Doubly Fed Induction Generator (DFIG)-based Wind Energy Conversion System (WECS) as shown in Fig. (**1.9**), the DFIG stator circuit is connected directly to the ac network at the point of common coupling (PCC) via a transformer, whereas rotor circuit is connected to the PCC via a back-to-back partial-scale voltage source converter (VSC) and the coupling transformer. The voltage source converter facilitates the variable speed operation of the DFIG [20, 21]. During the initial stage of introducing WECS to the electrical energy market, the disconnection of a wind turbine generators (WTGs) from the grid during a fault event at the ac network side was allowed to protect the wind turbine and the converter switches form any possible damages. Currently, the developed grid codes require WTGs to ride-through intermittent fault conditions to remain connected, thereby supporting the grid under such events. This will assure sustainable power delivery to the grid during faults and abnormal operating conditions. Improving DFIG fault ride through (FRT) capability can be achieved by implementing new control schemes for WECS to make them comply with the various grid codes that have been established by transmission system engineers as shown in the literature [22 - 26]; however this approach is only effective for new WECS installations. A cost effective approach to improve the FRT capability of the existing WECS is by connecting flexible ac transmission system (FACTS) device to the PCC [27]. A proper FACTS controller can also aid in overcoming some of the drawbacks of the DFIG (which include its high sensitivity to grid faults [28]). Several References can be found in the literatures on the low voltage ride through capability of the DFIG-based WECS [22 - 26]. Unfortunately, the problem of voltage swell has not been investigated thoroughly in the literature, despite the fact that its occurrence can cause a surge in the voltage at the point of common coupling and consequently violate the upper safety margin of the grid code requirements [29 - 31]. If not carefully mitigated, these faults may lead to voltage fluctuation across the terminal of the DFIG, high current within DFIG converters switches and voltage swell within the dc link capacitor of the converter stations.

As previously mentioned, application of the UPFC in power systems was investigated by many researchers [3, 6, 9, 11, 32 - 40]. However, studies concerning application of UPFC to WECS are limited. Moreover, majority of the existing studies limit the use of UPFC to smoothen the output power of fixed speed WECS during wind speed fluctuation to avoid system instability [32]. Following on from the limitations discussed above concerning UPFC, this chapter introduces new applications of the UPFC to improve the overall performance of DFIG-based WECS during fault events at the grid side. Simulation was carried out using the Simulink/Matlab software and the results obtained are analysed to examine the performance of WECS with and without the connection of UPFC.

Fig. (**2.6**) shows the investigated system, which consists of a wind farm of six DFIG-based wind turbines integrated with the ac grid at the point of common coupling (PCC), with each DFIG being rated at 1.5 MW. The DFIG rotor winding is connected to a bi-directional back to back voltage source converter, while the stator winding is connected directly to the ac network via a step up transformer. The wind turbines are connected to the grid via a 30 km transmission line (TL) and a step up transformer. The ac network is simulated as an ideal three-phase voltage source of constant frequency and voltage. Under normal operating conditions, the reactive power delivered by the DFIG is controlled at zero MVar to maintain a unity power factor connection at the PCC. In the proposed case study, the average wind speed is assumed to be 15 m/s and the generator's nominal speed is 1.2 pu, while nominal output active power of the turbine is 1.0 pu. The proposed UPFC is connected to the PCC bus.

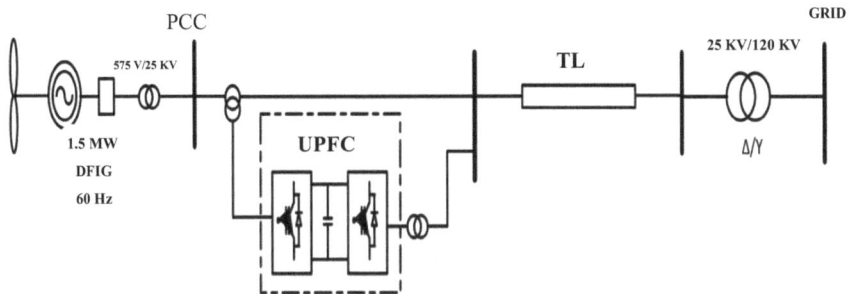

Fig. (2.6). Schematic of the system under study.

Proposed Control System

IGBTs based converters have been used in the proposed UPFC due to its advantages over GTOs, since IGBT has a higher switching frequency range (2-20 kHz) while GTO switching frequency is up to 1 kHz. The shunt and series converters of the UPFC are controlled using hysteresis current controller (HCC) and proportional Integral controller (PI), respectively as will be elaborated below. The widely used Park's transformation or the so-called dq transformation that was introduced by Robert H. Park in 1929 is used in the proposed controller. Park's transformation is a mathematical transformation that can be applied to any arbitrary three-phase system to calculate the equivalent two components in the dq reference frame that has the advantage of transforming the main component of waveform to a constant dc term under steady state conditions. The transformation equation is given below:

$$\begin{bmatrix} x_d \\ x_q \\ x_0 \end{bmatrix} = \sqrt{2/3} \begin{bmatrix} cos\theta_d & cos(\theta_d - 2\pi/3) & cos(\theta_d + 2\pi/3) \\ -sin\theta_d & -sin(\theta_d - 2\pi/3) & -sin(\theta_d + 2\pi/3) \\ 1/\sqrt{2} & 1/\sqrt{2} & 1/\sqrt{2} \end{bmatrix} \begin{bmatrix} x_a \\ x_b \\ x_c \end{bmatrix} \qquad (2.1)$$

Quantities in the dq reference frame can be transferred to three-phase using the inverse transformation as of (2.2) as:

$$\begin{bmatrix} x_a \\ x_b \\ x_c \end{bmatrix} = \sqrt{2/3} \begin{bmatrix} cos\theta_d & -sin\theta_d & 1/\sqrt{2} \\ cos(\theta_d - 2\pi/3) & -sin(\theta_d - 2\pi/3) & 1/\sqrt{2} \\ cos(\theta_d + 2\pi/3) & -sin(\theta_d + 2\pi/3) & 1/\sqrt{2} \end{bmatrix} \begin{bmatrix} x_d \\ x_q \\ x_0 \end{bmatrix} \qquad (2.2)$$

Where

$$\theta_d = \omega_d + \theta$$

ω_d is the angular velocity of signals to be transformed

θ is the angle of the initial state.

Hysteresis Current Controller

Hysteresis current control (HCC) method has the advantageous of being effective, simple and does not require complex processor to implement. The main concept of HCC is the generation of the converter switching pattern by comparing the actual values of the phase current with a fixed tolerance band around the reference value of the current associated with that phase. Fig. (**2.7**) shows the HCC scheme where the measured load current is maintained within the hysteresis band by controlling the switching state of the respective converter node voltage. However the other two phases can affect this type of band control and high frequency in the switching signal may rise as sequences of the interference between the phases (referred as inter-phases dependency). Inter-phases dependency can be reduced by applying the phase-locked loop technique to keep the converter switching at a fixed predetermined frequency level.

As shown in Fig. (**2.8**), deriving signals for the UPFC shunt converter switches are generated by comparing the values of the reference currents (Iabc*) with the three-phase line currents (Iabc) which are derived using Id* and Iq* references. The values of Id* and Iq* are generated based on the error values of the voltage across the dc-link capacitor (VDC) and the generator terminal voltage (Vg) by a traditional proportional integral (PI) controller. The value of Id* and Iq* is converted using the Park transformation (dq0-abc) to create the reference values of the current (Iabc*).

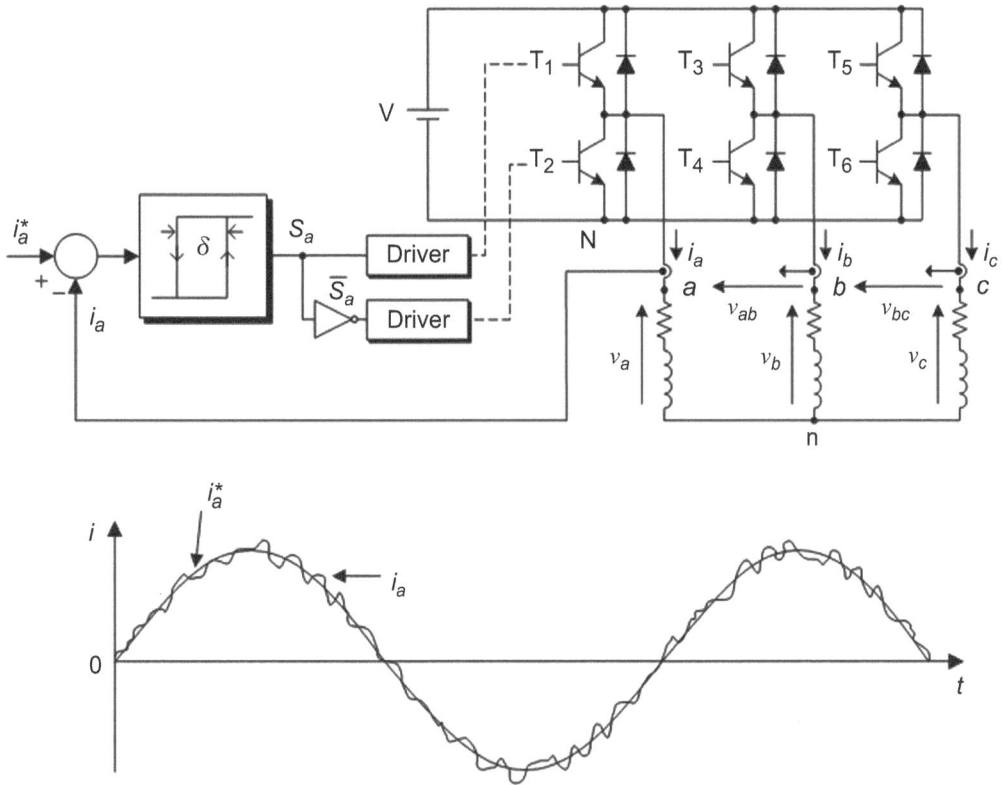

Fig. (2.7). Basic concept of Hysteresis current control for a three phase converter.

Proportional Integral Controller

The main role of the UPFC series converter is to control the active and reactive power flow through the transmission line by inserting a compensated voltage with controllable magnitude and phase in series with the transmission line. The a-b-c components at the point of common coupling (PCC) and the transmission line current are converted to the d-q reference frame using the Clarke-Park transformation technique. As shown in Fig. (**2.9**), the differences between the generator active and reactive powers and their pre-set values are used as input signals to the PI controllers which are tuned to generate the reference voltage values in d-q reference frame (Vd* and Vq*) required for controlling the series converter switches using pulse width nodulation (PWM). The active power P and reactive power Q in the d-q reference frame can be calculated as below:

$$P = V_d\, I_d + V_q\, I_q \tag{2.3}$$

$$Q = V_q\, I_d - V_d\, I_q \tag{2.4}$$

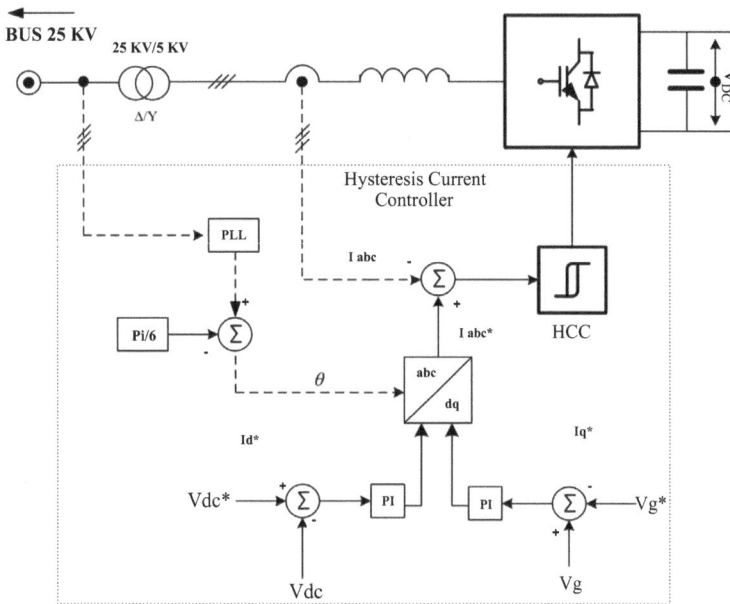

Fig. (2.8). Shunt converter HCC system.

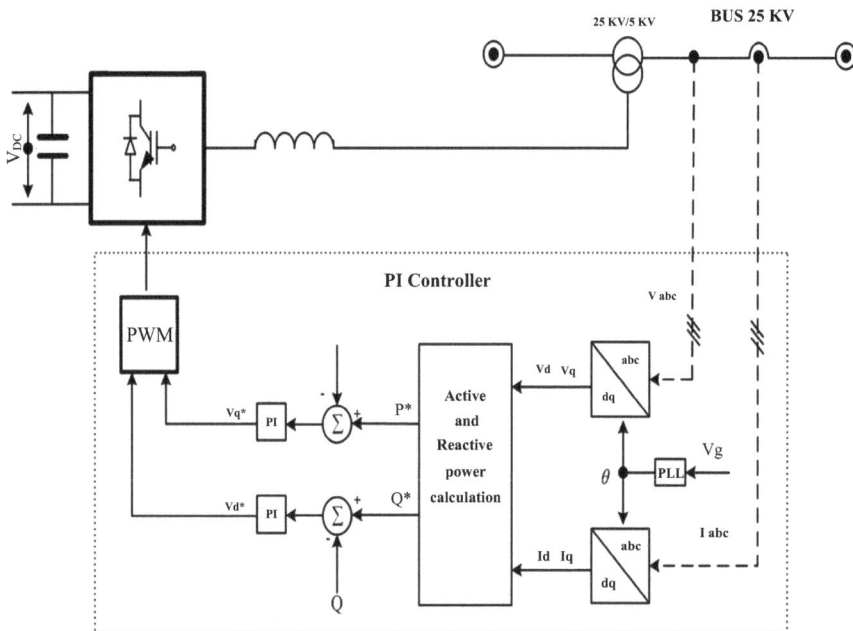

Fig. (2.9). Series converter control system.

Case Study 1: Low Voltage Ride Through

A significant voltage drop of 0.4 pu of the nominal value is assumed to take place at the grid side of the proposed system (Fig. **2.7**) at t = 6s, and this voltage drop is assumed to last for 5 cycles. The simulation results shown in Fig. (**2.10**), indicate that the voltage across the PCC will decrease to 0.43 pu when the UPFC is not connected. On the other hand, this voltage drop is regulated to a value of 0.87 pu by connecting the UPFC to the PCC bus. Based on the grid codes requirements, the regulated voltage (using UPFC) is within the safety margin of the investigated grid codes as can be seen in Fig. (**2.10**) [28].

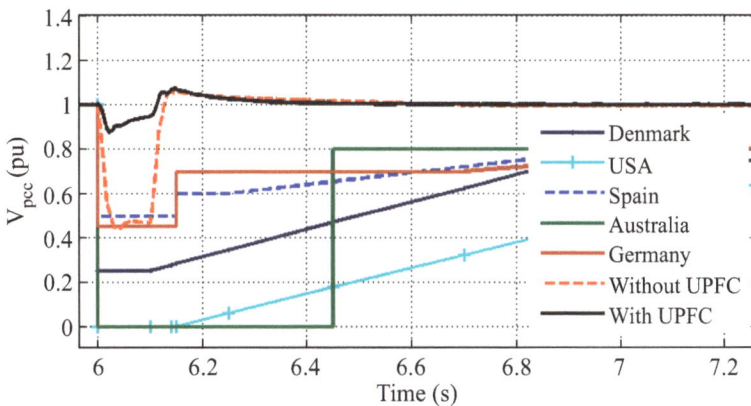

Fig. (2.10). Voltage at the PCC with and without UPFC during voltage sag.

Case Study 2: High Voltage Ride Through

Voltage swell occurs due to switching on a large capacitor bank or switching off a large load. A 1.4 pu voltage swell was simulated at the grid terminals and was assumed to last for 5 cycles, starting at t = 6s. (Fig. **2.11**) reveals that this fault causes the voltage at the PCC to increase by more than 30% without the connection of the UPFC, while this voltage is settled to a level less than 1.1 pu when the UPFC is connected to the PCC.

Case Study 3: Sub-Synchronous Resonance

Power systems are becoming larger and more interconnected, and as a consequence, the transient stability problem has become more serious. If the stability is lost, network collapse may occur with annihilating economical losses and severe power grid damages that may lead to overall blackout [41, 42]. Stability problems can occur due to the increase in the power demand with overloading the transmission lines; therefore, transmission line operators are required to increase the power transfer capability of the existing transmission lines. In this context, they have two options: the first one is to build an additional parallel transmission line, which is not a cost effective, especially for long transmission lines. The second option is to use a series capacitor as a partial compensation reactance to the transmission line, which has been extensively used as a very effective method to increase power transfer capability of transmission systems and improve their steady state and transient stability limits [43, 44]. Series capacitor is, however, not without problems, as it may cause sub-synchronous resonance (SSR) when the frequency of the generator mechanical shaft and the electrical frequency of the transmission system are added to the power frequency [45]. There are three ways in which a system and generator can interact with sub-synchronous effects: induction generator effect, torsional interaction and transient torques [45]. SSR due to transient torque can be developed if the electrical resonant frequency of the network is complementary to any of the natural torsional oscillating frequencies of the turbine-generator shaft during system disturbance events. The electric resonance of the transmission system and the torsional oscillations of the mass-spring system of the turbine generator will be mutually excited and might get augmented, causing serious shaft oscillations and shaft fatigue and possibly damage and failure [46]. The first two shaft failures due to SSR occurred at the Mohave power station in 1970 and 1971, respectively [47 - 50]. It is important to investigate the sub-synchronous resonance when planning to include series capacitors for new or existing transmission lines. Extensive research was undertaken to increase the damping of the torsional mode and many countermeasures were suggested to dampen the SSR. Some suggested solutions include use of the synchronous-machine-based Energy Storage System (ESS) [51], static var compensator (SVC) [52, 53], superconducting magnetic energy storage (SMES) unit [54 - 57], Static Synchronous Compensator (STATCOM) [58], shunt reactor controller [59, 60], Thyristor-controlled dynamic resistance braking [48, 61], excitation control of synchronous generator [62, 63] and gate controlled series capacitors [64, 65].

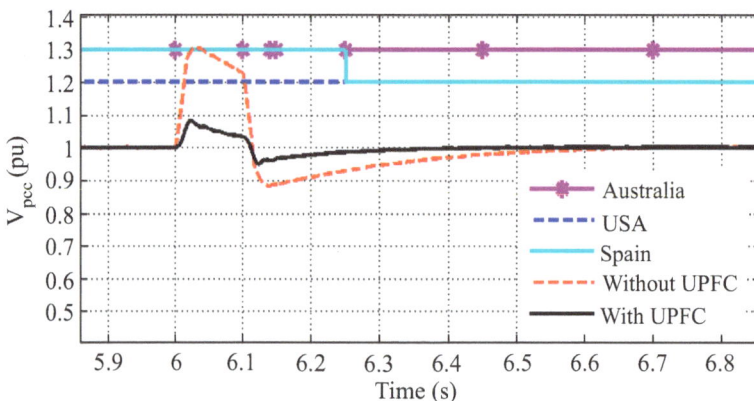

Fig. (2.11). Voltage at the PCC with and without UPFC during voltage swell.

The case study presented in this section investigates the use of UPFC to enhance transient stability and to dampen the SSR of a steam turbine-generator connected to a large interconnected ac grid via a series capacitor compensated transmission line.

The system shown in Fig. (**2.6**) is modified to the one shown in Fig. (**2.12**). This system consists of a synchronous generator and wind farm that includes 6 wind turbines of type D connected to the grid (simulated as infinite bus via a Y/Δ step down transformer) and two parallel transmission lines: one of which is series compensated. The UPFC is connected to bus-1 between the compensated transmission lines and the generators to provide adequate damping for the turbine generator set.

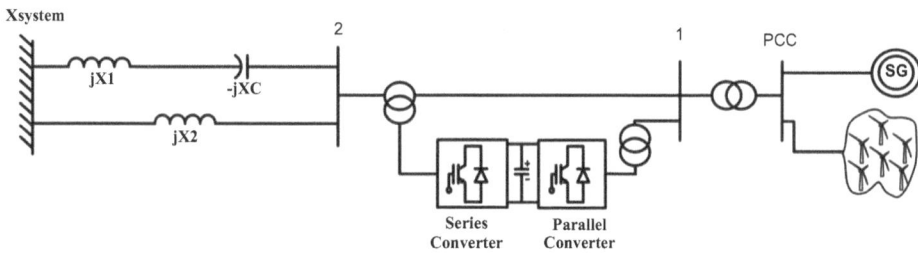

Fig. (2.12). The investigated system for SSR study.

The shaft system of the turbine generator set consists of four masses; a high-pressure turbine (HP), a low-pressure turbine (LP), a generator rotor (Gen) and an exciter (Ex). The system is simulated with inclusion of all non-linearities, such as exciter ceiling voltage limit. A three phase short-circuit fault is simulated at bus-2 at t = 12 s and is assumed to be cleared at t = 12.035 s. The compensation degree (X_c/X_l) is assumed to be 55%. Figs. (**2.13** to **2.21**) show the dynamic response of the studied system with and without the UPFC. The effect of the SSR and the UPFC controller is investigated through time domain waveforms of various system variables such as the synchronous generator speed deviation, deviation in torsional torque induced on the shaft sections between the high-pressure, low-pressure turbines (T_{HP-LP}) and the low-pressure turbine and generator (T_{Gen-LP}), the DFIG electromechanical torque (T_{em}), the PCC voltage, the DFIG converter dc voltage and the DFIG speed. Without the UPFC and due to lack of damping, the system is unstable as evinced by the high torsional forces induced in the generator mechanical shafts and the significant increment in the synchronous generator shaft speeds, as can be seen in Figures below.

When the UPFC is connected, damping of the synchronous generator and the DFIG is greatly enhanced and the stability margin can be extended. It can also be shown that using a UPFC reduces the high torsional forces on the turbine-generator shaft sections approximately to the normal steady state values; it also decreases the settling time substantially. Connecting the UPFC also reduces the generator shaft speed oscillations and maintains the speed at the nominal value, as clearly shown in Fig. (**2.21**).

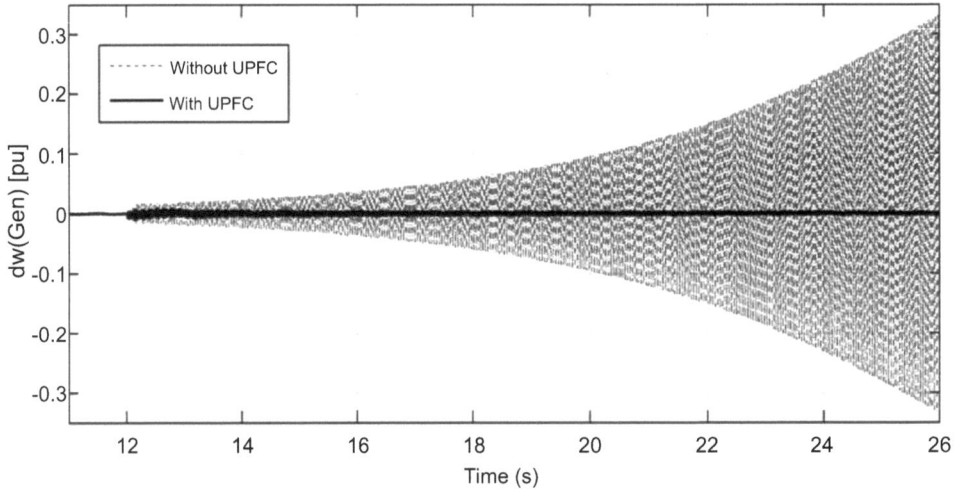

Fig. (2.13). Deviation of the synchronous generator speed with and without UPFC.

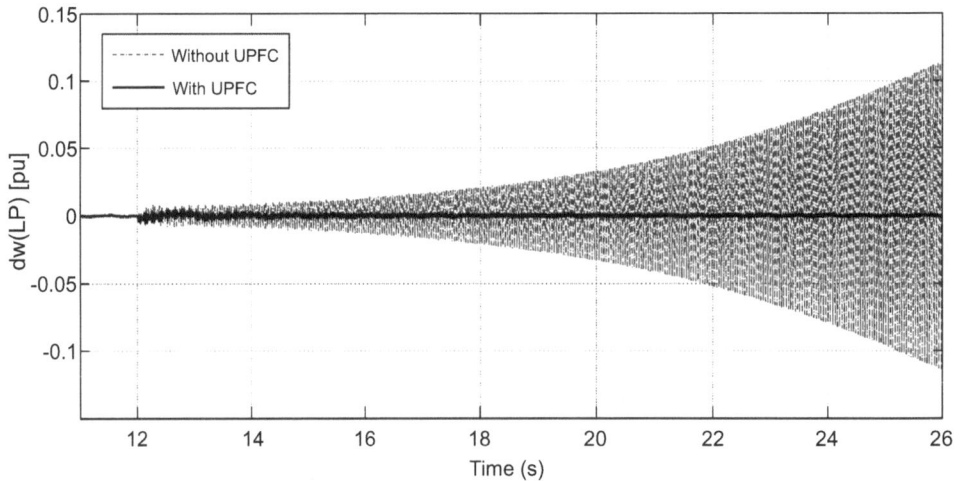

Fig. (2.14). LP speed deviation with and without UPFC.

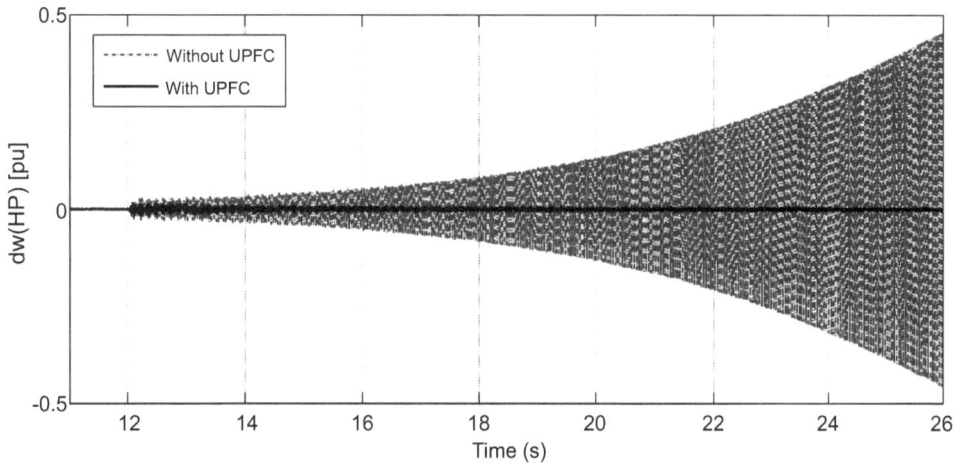

Fig. (2.15). HP speed deviation with and without UPFC.

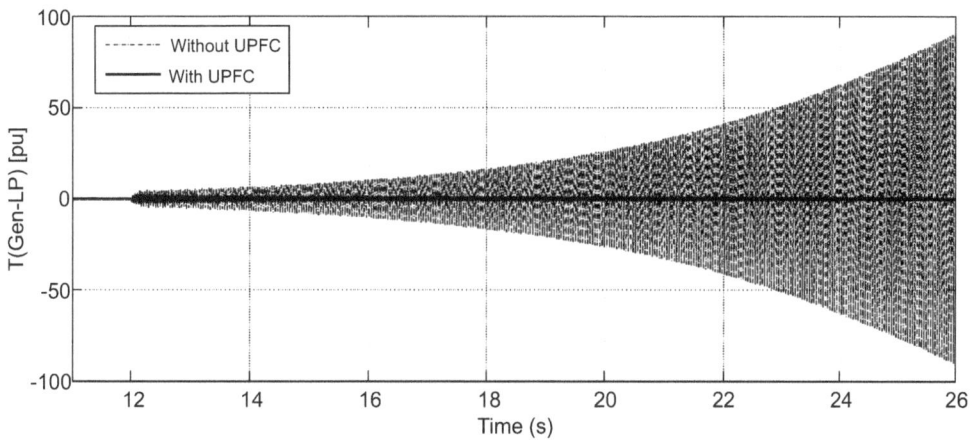

Fig. (2.16). LP to Gen Torque deviation with and without UPFC.

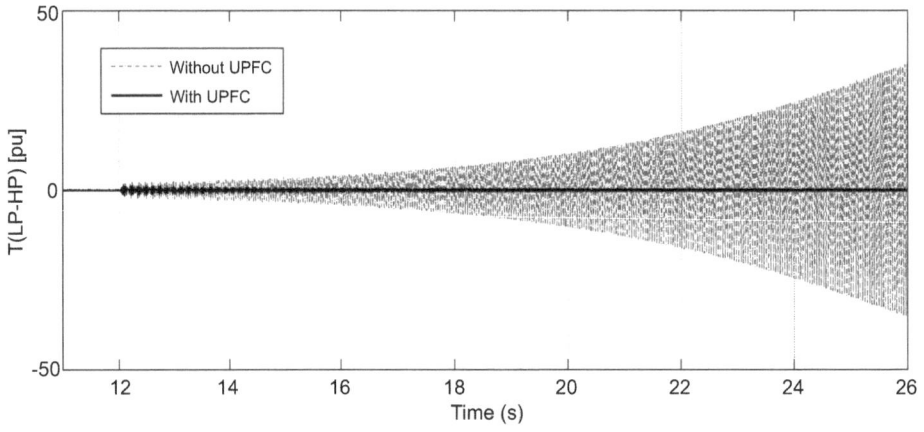

Fig. (2.17). HP to LP Torque deviation with and without UPFC.

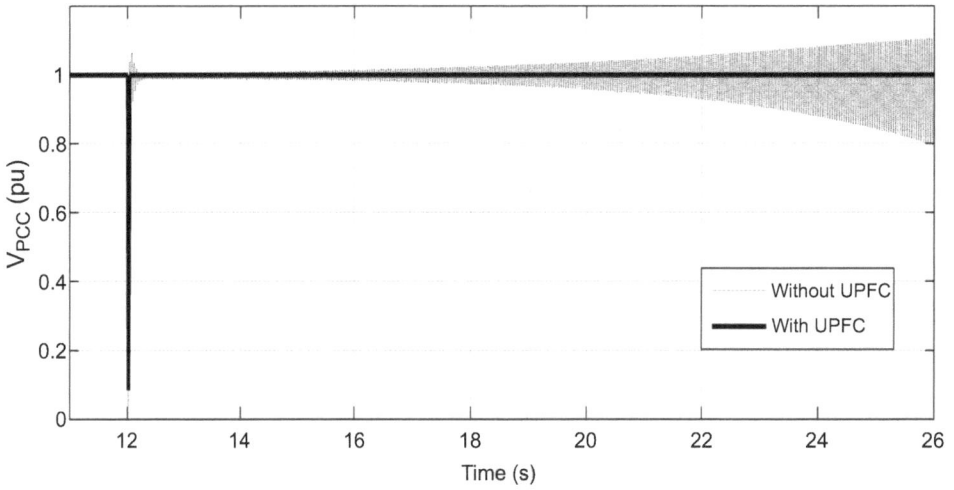

Fig. (2.18). The PCC voltage with and without UPFC.

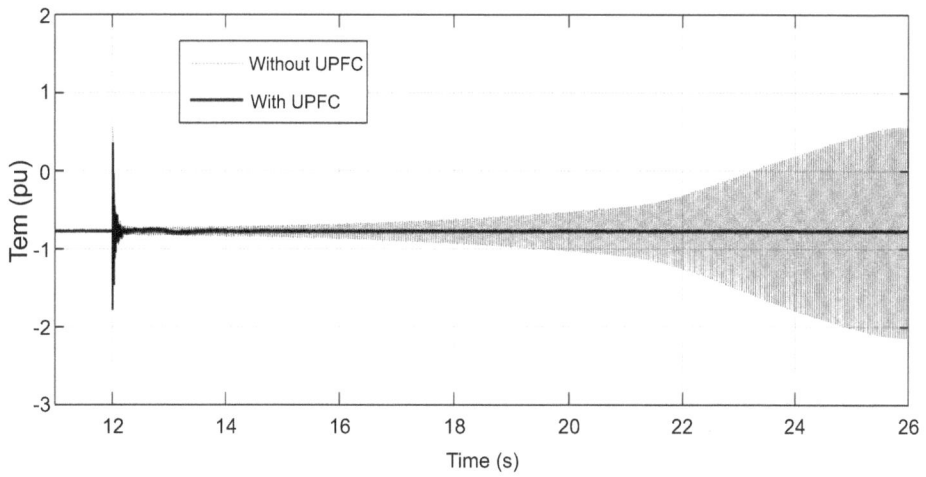

Fig. (2.19). DFIG electromechanical with and without UPFC.

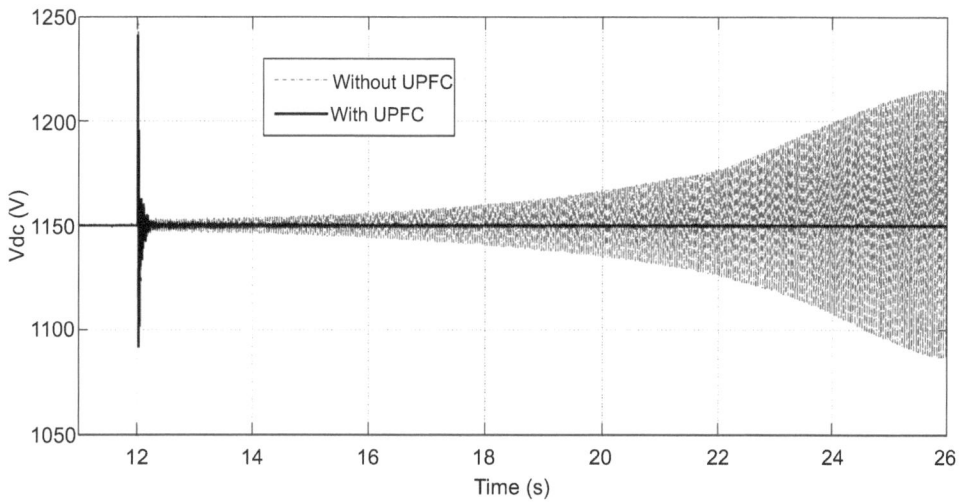

Fig. (2.20). DFIG V_{dc} response with and without UPFC.

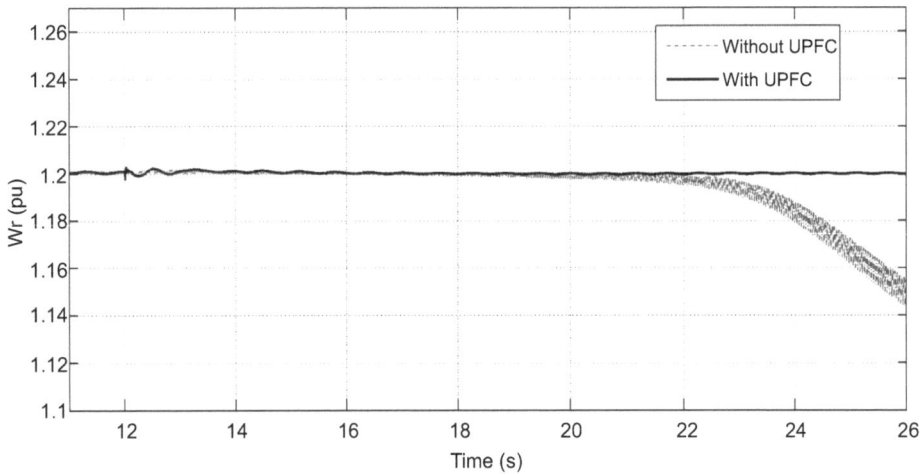

Fig. (2.21). DFIG speed with and without UPFC.

Application of UPFC to Improve the Performance of DFIG During Converter Station Faults

While there are some studies about the effect of internal converter station faults (such as misfire and fire-through) on the performance of high-voltage direct-current systems [66, 67], only limited attention has been given to investigate the impact of such faults on the overall performance of the DFIG-based WECS and the compliance of the DFIG with the recent developed grid codes during such faults [68, 69]. Misfire is the failure of the converter switch in taking over conduction at a scheduled programmed conducting period. On the other hand, fire through is the failure of the converter switch to block during a scheduled, non-conducting period [70]. These internal converter faults are caused by various malfunctions in the control and firing equipment [71]. Industrial survey about failure distribution in the converter stations that interface HVDC lines with ac network shows that converter faults due to malfunctions within the control circuit represent about 53.1% while about 37.9% of the converter faults are attributed to converter power parts (Fig. **2.22**) [72, 73]. Some of the converter faults are self-clearing if the causes are of transient nature; however, they can still have a detrimental impact on the system, particularly when they occur within the inverter station rather than the rectifier station [74]. The use of an insulated-gate bipolar transistor (IGBT) in both DFIG converters is preferred due to its advantages when compared with gate turn-off thyristor (GTO) [75]. When a malfunction occurs on the IGBT-based converter station, it can cause catastrophic breakdown to the device, if the fault remains undetected [76].

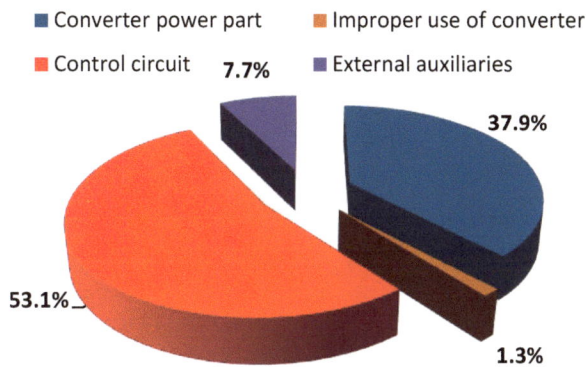

Fig. (2.22). Distribution of failure types in the converters [74].

Faults within Rotor Side Converter

In this section, the DFIG dynamic performance under misfire and fire-through intermittent faults within the RSC is investigated. The two aforementioned faults are assumed to take place on one switch within the RSC at 4.1 s and cleared at 4.15 s. Fig. (**2.20**) shows the dynamic performance of the DFIG during the two switch faults. As can be seen in Fig. (**2.23a**), the fire-through fault causes the voltage at the PCC to drop to 0.45 pu, while the misfire fault impact on the voltage at the PCC is insignificant. During steady state condition, DFIG delivers 1 pu active power and 0 pu reactive power to the ac network, as shown in Fig. (**2.23b, c**). However, when the fire-through fault takes place in one of the switches, a significant drop in the delivered active power will occur and the reactive power will be absorbed from the grid. On the other hand, the impact on DFIG power is insignificant when misfire fault takes place. The shaft speed behaviour can be seen in Fig. (**2.23d**), where the shaft speed experiences oscillations and overshooting during the two faults; the impact is more pronounced in the case of fire-through fault. As can be observed in Fig. (**2.23e**), the fire through fault will cause short circuit across the dc capacitor when switch S4 takes over conduction along with the faulty switch (S1), and as a result, the voltage across the dc capacitor link will drop to zero level. The impact of misfire on this voltage is insignificant. The voltage across the capacitor is designed with safety requirement set by the designers; in [77] this safety level is between 0.25 pu and 1.25 pu. If this safety level is violated, the converter station protection system must act to disconnect the converter to avoid possible damages to the switches.

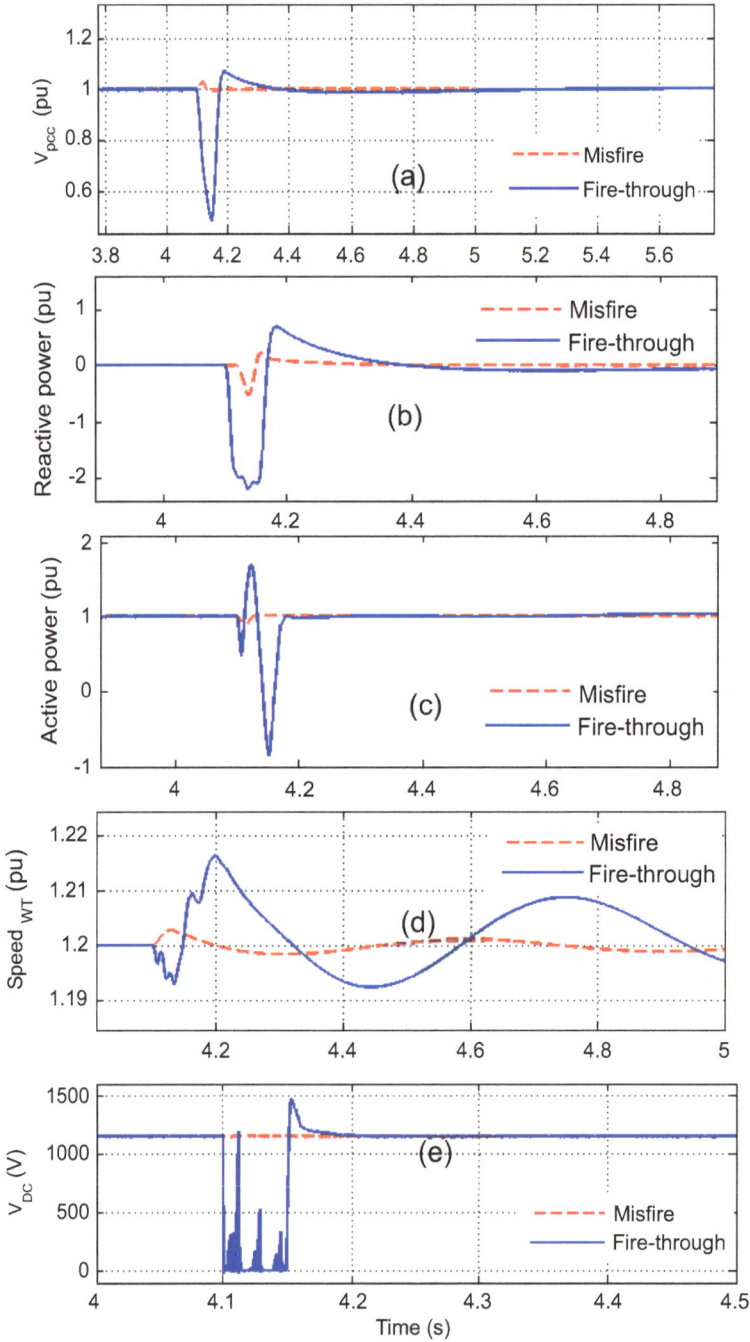

Fig. (2.23). DFIG performance during fire-through and misfire within the rotor side converter: (a) PCC voltage, (b) reactive power, (c) active power (d) shaft speed, and (e) dc link Voltage of DFIG.

The impact of misfire and fire-through faults on converter terminal voltages is shown in Fig. (**2.24**). Spikes are introduced to the converter terminal voltages when switch S1 of the RSC experiences misfire. However, the fire-through fault on switch S1 causes short circuit across the converter terminals, causing the voltage to drop to zero level during the fault.

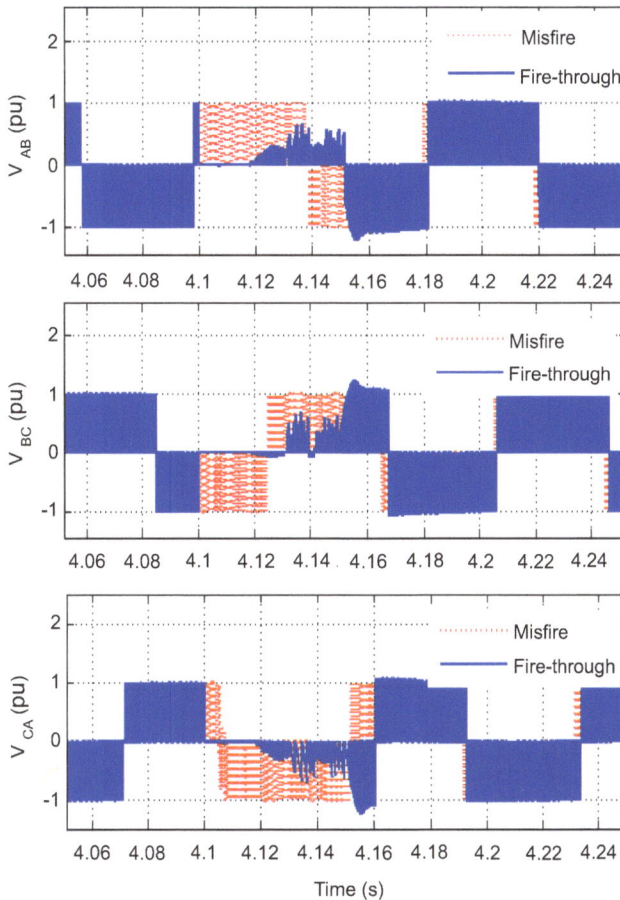

Fig. (2.24). Voltage change across RSC terminals during fire-through and misfire in S1 within RSC.

Faults within the Grid Side Converter

In this case study, the GSC switch S1 is assumed to experience misfire and fire-through within the period 4.1 s to 4.15 s. As shown in Fig. (**2.25a**), due to the fire-through fault, the voltage at the PCC decreases by 0.55 pu, whereas the impact of misfire fault is negligible. During the event of fire-through the DFIG absorbs a significant amount of reactive power from the ac network and the real power produced by the DFIG is significantly decreased, as shown in Fig. (**2.25b**) and Fig. (**2.25c**), respectively.

In contrast, both the reactive and active powers are slightly affected during the misfire fault event. The DFIG power drops cause acceleration to the shaft speed to compensate for the power imbalance during the fire-through fault, as can be shown in Fig. (**2.25d**), with no impact due to the misfire fault. It can be seen from Fig. (**2.25e**), the fire-through fault causes the dc voltage across the capacitor to drop to zero level; while in the case of misfire, the dc voltage experiences slight fluctuation. The fire-through fault causes a voltage collapse to zero level across the converter terminals, as shown in Fig. (**2.26**). Similar to RSC, the impact on the converter terminal voltages is negligible when the misfire fault takes place on switch S1 of the GSC.

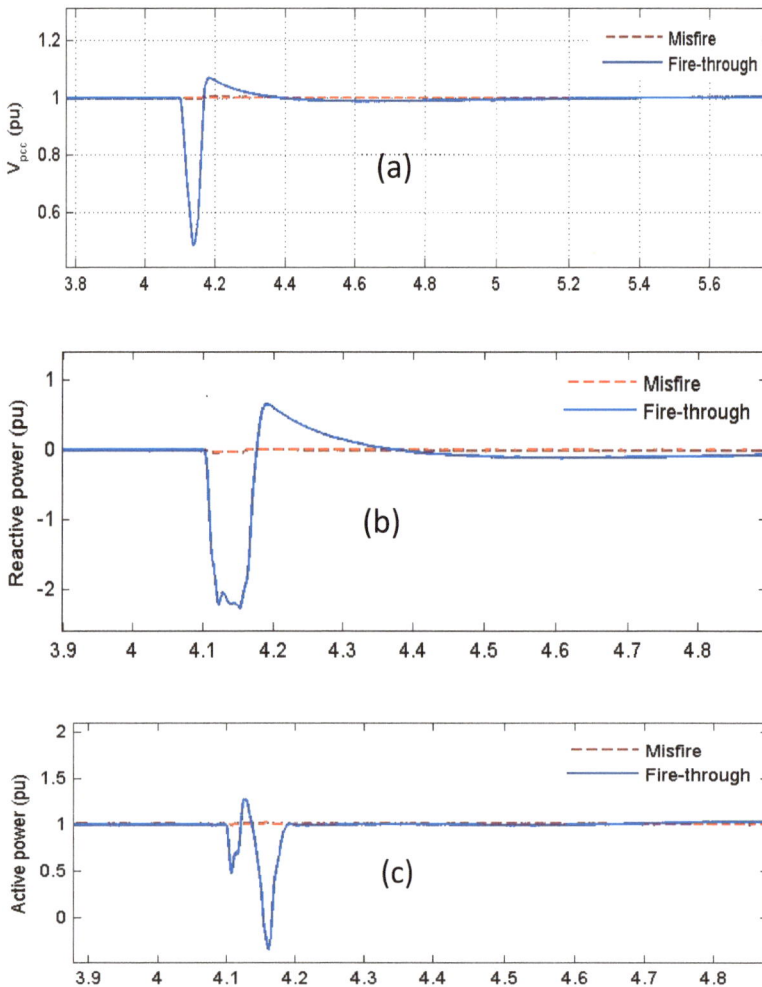

Fig. 2.25 contd.....

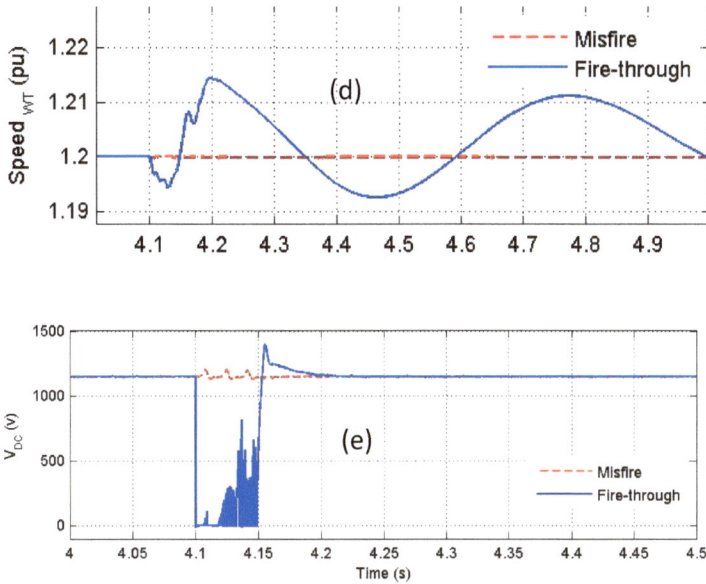

Fig. (2.25). DFIG performance during fire-through and misfire within the grid side converter; (**a**) PCC voltage, (**b**) reactive power, (**c**) active power (**d**) shaft speed, and (**e**) dc link voltage of DFIG.

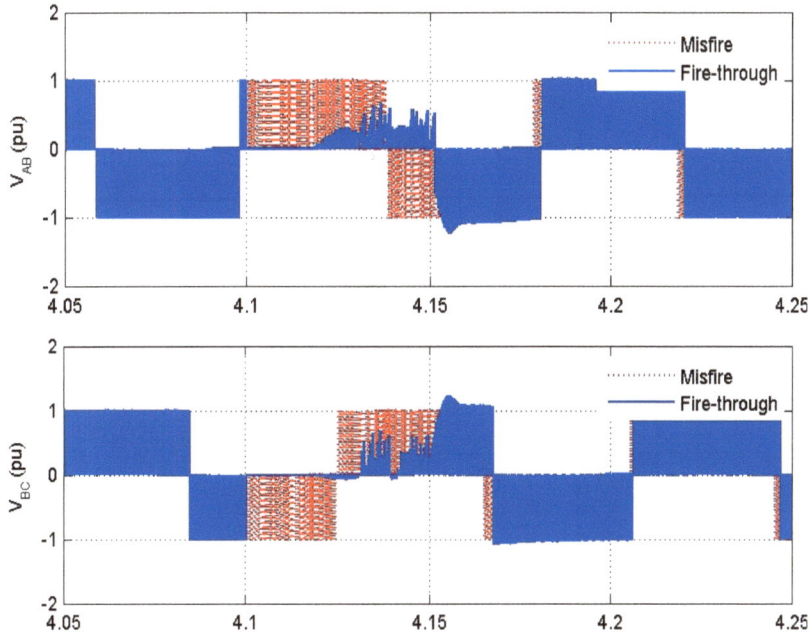

Fig. 2.26 contd.....

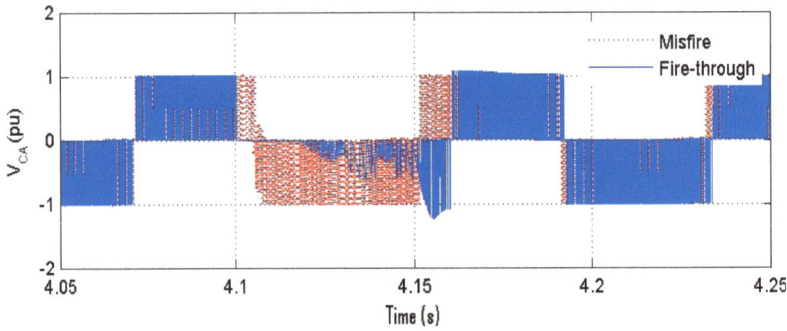

Fig. (2.26). Voltage change across GSC terminals during fire-through and misfire in S1 within GSC.

Impact of UPFC During Converter Station Faults

To mitigate the adverse impacts of fire-through faults on the DFIG performance, the UPFC is connected to the PCC, as shown in Fig. (**2.27**). As the misfire fault has insignificant impact, it is not included in this section. Fire-through fault studied in the previous section is re-investigated but with connection of the UPFC.

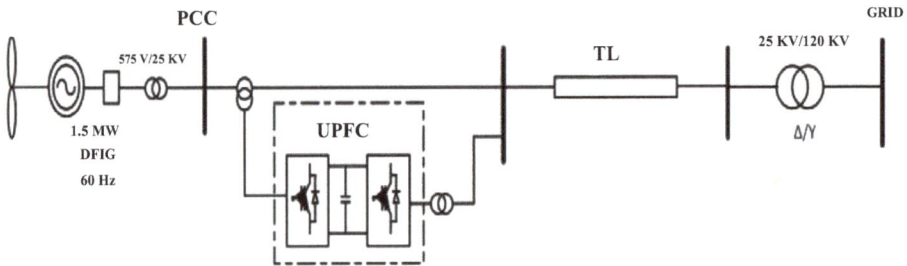

Fig. (2.27). Proposed System with UPFC.

Fire-through Fault

When a fire-through fault takes place within the RSC S1, the voltage at the PCC drops by 0.5 pu when the UPFC is not connected. This voltage drop is regulated to 0.69 pu by connecting the UPFC to the PCC bus, as shown in Fig. (**2.28a**). As discussed in the previous section, without any compensation, the generated active power drops and the shaft speed accelerates, as shown in Fig. (**2.28b, c**). With the connection of the UPFC, the drop in DFIG generated power is reduced, leading to a reduction in the speed overshooting and settling time.

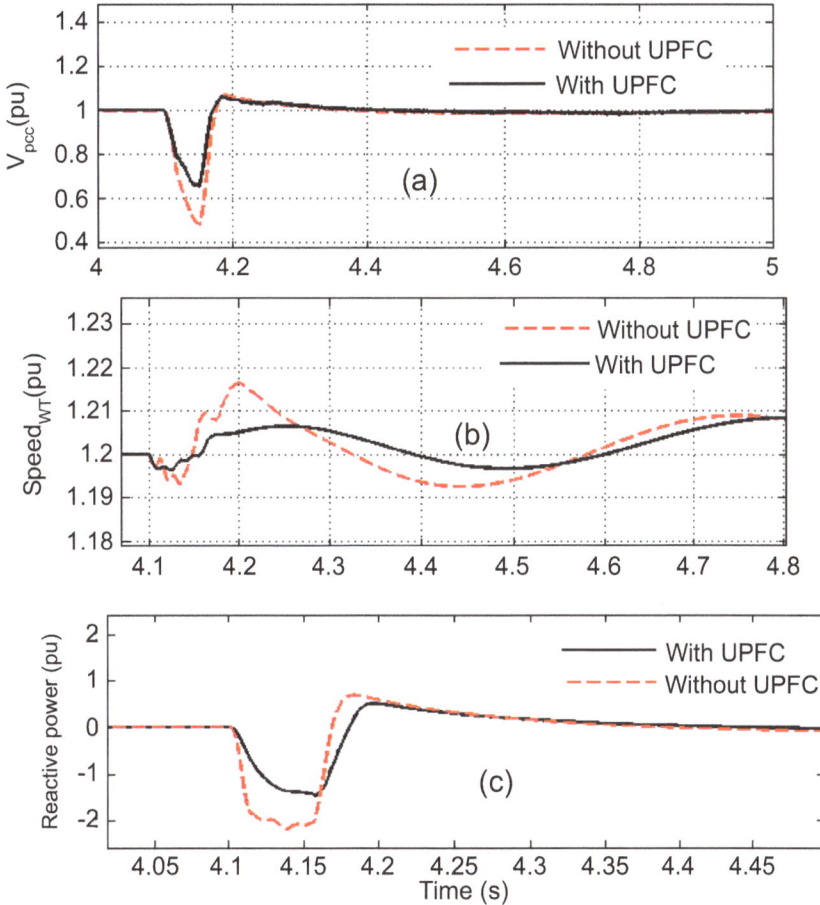

Fig. (2.28). Effect of RSC Fire-through on DFIG dynamic performance with and without UPFC: (**a**) Voltage at PCC. (**b**) Reactive power, (**c**) Shaft speed.

When the fire-through takes place within the Grid side converter, the same trend can be observed when the UPFC is connected to the PCC bus. Fig. (**2.29a**) shows that the UPFC can rectify the voltage drop at the PCC from 0.47 pu to 0.97 pu. This regulation is critical, since without this compensation, the DFIG must be disconnected according to some grid codes such as, the Spanish grid code, which specifies a 0.5 pu maximum voltage sag at the PCC to maintain the wind turbine connection to the grid. The generator shaft speed is slightly retarded through the fault duration, and it experiences overshooting upon recovery of the fault and settles at a lower steady state level than the pre-set level (1.2 pu), as shown in Fig. (**2.29c**).

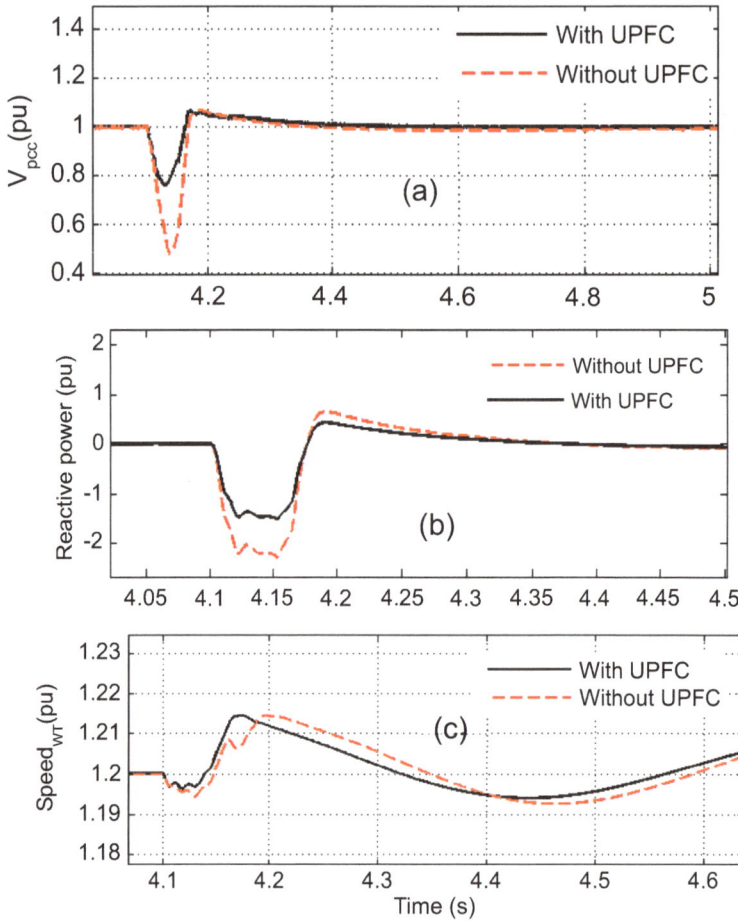

Fig. (2.29). Effect of GSC Fire-through on DFIG dynamic performance without and with UPFC: (a) Voltage at PCC. (b) Reactive power, (c) Shaft speed. (d) dc-link Voltage.

SUMMARY

This chapter presents case studies for various applications of UPFC in the DFIG-based WECS. New applications for UPFC have been proposed to improve the overall performance of a DFIG-based WECS during voltage sag and voltage swell event at the grid side. Results from simulation work show that the proposed UPFC succeeded to improve the dynamic performance of the studied DFIG-based WECS during various fault events. Moreover, the chapter has investigated the application of the UPFC to stabilize multi-mode torsional oscillations of sub-synchronous resonance. The simulation results indicated that the proposed UPFC is very effective in damping all SSR modes of the investigated system and in minimizing the potential disconnection of the wind farm during the studied faults. Application of the UPFC to improve the performance of DFIG during converter station faults has been investigated. Simulation results of the converter station faults showed that fire-through can have severe impacts on the dynamic performance of the DFIG, and this in turn can result in severe damages to the WTG. However, with the introduction of UPFC the system performance improves and the impacts of the fault are mitigated. The analysis showed that misfire has minor effects on the DFIG performance.

REFERENCES

[1]   A. Berizzi, M. Delfanti, P. Marannino, M.S. Pasquadibisceglie, and A. Silvestri, "Enhanced Security-Constrained OPF With FACTS Devices", *IEEE Trans. Power Syst.,* vol. 20, pp. 1597-1605, 2005. [http://dx.doi.org/10.1109/TPWRS.2005.852125]

[2]   N. Hingorani, and L. Gyugyi, "Combined Compensators: Unified Power Flow Controller (UPFC) and Interline Power Flow Controller (IPFC)", In: Understanding FACTS:Concepts and Technology of Flexible ac Transmission Systems, ed: Wiley-IEEE Press, 2000, pp. 297-352.

[3]   A. Rajabi-Ghahnavieh, M. Fotuhi-Firuzabad, M. Shahidehpour, and R. Feuillet, "UPFC for Enhancing Power System Reliability", *Power Delivery, IEEE Transactions on,* vol. 25, pp. 2881-2890, 2010. [http://dx.doi.org/10.1109/TPWRD.2010.2051822]

[4]   H. Akagi, E.H. Watanabe, and M. Aredes, "Instantaneous Power Theory and Applications to Power Conditioning", ed: Wiley-IEEE Press. [http://dx.doi.org/10.1002/9781119307181]

[5]   L. G. c. E.-H. Narain G. Hingorani, consulting editor., Understanding FACTS : concepts and technology of flexible ac transmission systems New York : IEEE Press 2000.

[6]   E. Fuchs, and M.A. Masoum, *Power Quality in Power Systems and Electrical Machines.* Elsevier Science, 2011.

[7]   N. Talebi, and A. Akbarzadeh, "Damping of Low Frequency Oscillations in power systems with neuro-fuzzy UPFC controller", *Environment and Electrical Engineering (EEEIC), 2011 10th International Conference on,* 2011pp. 1-4 [http://dx.doi.org/10.1109/EEEIC.2011.5874855]

[8]   C. Chia-Chi, and T. Hung-Chi, "Application of Lyapunov-based adaptive neural network upfc damping controllers for transient stability enhancement", *Power and Energy Society General Meeting - Conversion and Delivery of Electrical Energy in the 21st Century, 2008 IEEE,* 2008, pp. 1-6 [http://dx.doi.org/10.1109/PES.2008.4596580]

[9]   J. Guo, M.L. Crow, and J. Sarangapani, "An Improved UPFC Control for Oscillation Damping", *Power Systems, IEEE Transactions on,* vol. 24, pp. 288-296, 2009. [http://dx.doi.org/10.1109/TPWRS.2008.2008676]

[10]   P.K. Dash, S. Morris, and S. Mishra, "Design of a nonlinear variable-gain fuzzy controller for FACTS devices", *IEEE Trans. Contr. Syst. Technol.,* vol. 12, pp. 428-438, 2004. [http://dx.doi.org/10.1109/TCST.2004.824332]

[11]   Y.L. Kang, G.B. Shresta, and T.T. Lie, "Application of an NLPID controller on a UPFC to improve transient stability of a power system", *Generation, Transmission and Distribution, IEE Proceedings,* vol. 148, 2001, pp. 523-529 [http://dx.doi.org/10.1049/ip-gtd:20010526]

[12]   R.K. Khadanga, and S. Panda, "Gravitational search algorithm for Unified Power Flow Controller based damping controller design", *Energy, Automation, and Signal (ICEAS), 2011 International Conference on,* 2011, pp. 1-6 [http://dx.doi.org/10.1109/ICEAS.2011.6147130]

[13]   S. Ray, and G.K. Venayagamoorthy, "Wide-Area Signal-Based OptimalNeurocontroller for a UPFC", *Power Delivery, IEEE Transactions on,* vol. 23, pp. 1597-1605, 2008. [http://dx.doi.org/10.1109/TPWRD.2007.916111]

[14]   L. Liming, Z. Pengcheng, K. Yong, and C. Jian, "Power-Flow Control Performance Analysis of a Unified Power-Flow Controller in a Novel Control Scheme", *Power Delivery, IEEE Transactions on,* vol. 22, pp. 1613-1619, 2007. [http://dx.doi.org/10.1109/TPWRD.2006.886799]

[15]   S. Kannan, S. Jayaram, and M.M. Salama, "Real and reactive power coordination for a unified power flow controller", *Power Systems, IEEE Transactions on,* vol. 19, pp. 1454-1461, 2004. [http://dx.doi.org/10.1109/TPWRS.2004.831690]

[16]   P.K. Dash, S. Mishra, and G. Panda, "A radial basis function neural network controller for UPFC", *Power Systems, IEEE Transactions on,* vol. 15, pp. 1293-1299, 2000. [http://dx.doi.org/10.1109/59.898104]

[17]   M.J. Kumar, S.S. Dash, A.S. Immanuvel, and R. Prasanna, "Comparison of FBLC (Feed-Back Linearisation) and PI-Controller for UPFC to enhance transient stability", *Computer, Communication and Electrical Technology (ICCCET), 2011 International Conference on,* 2011, pp. 376-381

[18]   H.I. Shaheen, G.I. Rashed, and S.J. Cheng, "Design of new nonlinear optimal predictive controller for Unified Power Flow Controller", *Power and Energy Society General Meeting - Conversion and Delivery of Electrical Energy in the 21st Century, 2008 IEEE,* 2008, pp. 1-10 [http://dx.doi.org/10.1109/PES.2008.4596242]

[19]   F. Shalchi, H. Shayeghi, and H.A. Shayanfar, "Robust control design for UPFC to improve damping of oscillation in distribution system by H<inf>2</inf> method", *in Electrical Power Distribution Networks (EPDC), 2011 16th Conference on* , 2011, pp. 1-10

[20]   J.M. Carrasco, L.G. Franquelo, J.T. Bialasiewicz, E. Galvan, R.C. Guisado, and M.A. Prats, "Power-Electronic Systems for the Grid Integration of Renewable Energy Sources: A Survey", *IEEE Trans. Ind. Electron.,* vol. 53, pp. 1002-1016, 2006. [http://dx.doi.org/10.1109/TIE.2006.878356]

[21]   T. Ackerman, *Wind Power in Power System.* John Wiley and Sons Ltd: West Sussex, 2005. [http://dx.doi.org/10.1002/0470012684]

[22]   S. Seman, J. Niiranen, and A. Arkkio, "Ride-Through Analysis of Doubly Fed Induction Wind-Power Generator Under Unsymmetrical Network Disturbance", *IEEE Trans. Power Syst.,* vol. 21, pp. 1782-1789, 2006. [http://dx.doi.org/10.1109/TPWRS.2006.882471]

[23]   J. Lopez, E. Gubia, E. Olea, J. Ruiz, and L. Marroyo, "Ride Through of Wind Turbines With Doubly Fed Induction Generator Under Symmetrical Voltage Dips", *IEEE Trans. Ind. Electron.,* vol. 56, pp. 4246-4254, 2009. [http://dx.doi.org/10.1109/TIE.2009.2028447]

[24]   M. Mohseni, S.M. Islam, and M.A. Masoum, "Impacts of Symmetrical and Asymmetrical Voltage Sags on DFIG-Based Wind Turbines Considering Phase-Angle Jump, Voltage Recovery, and Sag Parameters", *IEEE Trans. Power Electron.,* vol. 26, pp. 1587-1598, 2011. [http://dx.doi.org/10.1109/TPEL.2010.2087771]

[25]   Y. Xiangwu, G. Venkataramanan, P.S. Flannery, W. Yang, D. Qing, and Z. Bo, "Voltage-Sag Tolerance of DFIG Wind Turbine With a Series Grid Side Passive-Impedance Network", *IEEE Trans. Energ. Convers.,* vol. 25, pp. 1048-1056, 2010. [http://dx.doi.org/10.1109/TEC.2010.2054097]

[26]   S. Hu, X. Lin, Y. Kang, and X. Zou, "An Improved Low-Voltage Ride-Through Control Strategy of Doubly Fed Induction Generator During Grid Faults", *IEEE Trans. Power Electron.,* vol. 26, pp. 3653-3665, 2011.

[http://dx.doi.org/10.1109/TPEL.2011.2161776]

[27]   S.M. Muyeen, R. Takahashi, T. Murata, and J. Tamura, "A Variable Speed Wind Turbine Control Strategy to Meet Wind Farm Grid Code Requirements", *Power Systems, IEEE Transactions on,* vol. 25, pp. 331-340, 2010.
[http://dx.doi.org/10.1109/TPWRS.2009.2030421]

[28]   V. Ahkmatov, " Analysis of dynamic behaviour of power systems with large amount of wind power", Available: http://www.dtu.dk/upload/centre/cet/projekter/99-05/05-va-thesis.pdf

[29]   A.M. Yunus, M.A. Masoum, and A. Abu-Siada, "Application of SMES to Enhance the Dynamic Performance of DFIG During Voltage Sag and Swell", *Applied Superconductivity, IEEE Transactions on,* vol. 22, pp. 5702009-5702009, 2012.
[http://dx.doi.org/10.1109/TASC.2012.2191769]

[30]   A.M. Yunus, A.A. Siada, and M.A. Masoum, "Application of SMES Unit to Improve the High Voltage Ride Through Capabillty of DFIG-Grid Connected during Voltage Swell", *IEEE Power and Energy Society Innovative Smart Grid Technologies Conference (ISGT 2011),* 2011 Perth, Australia

[31]   Alt, x, M. n, Go, O. ksu, R. Teodorescu, *et al.,* "Overview of recent grid codes for wind power integration," in 12th International Conference on Optimization of Electrical and Electronic Equipment (OPTIM), 2010 2010, pp. 1152-1160.

[32]   W. Li, L. Hao-Wen, and W. Cheng-Tai, "Stability Analysis of an Integrated Offshore Wind and Seashore Wave Farm Fed to a Power Grid Using a Unified Power Flow Controller", *Power Systems, IEEE Transactions on,* vol. 28, pp. 2211-2221, 2013.
[http://dx.doi.org/10.1109/TPWRS.2013.2237928]

[33]   L. Gyugyi, "A unified flow control concept for flexible ac transmission systems", *ac and dc Power Transmission, 1991., International Conference on,* 1991pp. 19-26

[34]   E. Gholipour, and S. Saadate, "Improving of transient stability of power systems using UPFC", *Power Delivery, IEEE Transactions on,* vol. 20, pp. 1677-1682, 2005.
[http://dx.doi.org/10.1109/TPWRD.2005.846354]

[35]   S. Shojaeian, J. Soltani, and G.A. Markadeh, "Damping of Low Frequency Oscillations of Multi-Machine Multi-UPFC Power Systems, Based on Adaptive Input-Output Feedback Linearization Control", *Power Systems, IEEE Transactions on,* vol. 27, pp. 1831-1840, 2012.
[http://dx.doi.org/10.1109/TPWRS.2012.2194313]

[36]   N. Tambey, and M.L. Kothari, "Damping of power system oscillations with unified power flow controller (UPFC)", *Generation, Transmission and Distribution, IEE Proceedings,* vol. 150, 2003pp. 129-140
[http://dx.doi.org/10.1049/ip-gtd:20030114]

[37]   D. Thukaram, L. Jenkins, and K. Visakha, "Improvement of system security with unified-power-flow controller at suitable locations under network contingencies of interconnected systems", *Generation, Transmission and Distribution, IEE Proceedings,* vol. 152, 2005pp. 382-390
[http://dx.doi.org/10.1049/ip-gtd:20045235]

[38]   R.L. Vasquez-Arnez, and L.C. Zanetta, "Compensation strategy of autotransformers and parallel lines performance, assisted by the UPFC", *Power Delivery, IEEE Transactions on,* vol. 20, pp. 1550-1557, 2005.
[http://dx.doi.org/10.1109/TPWRD.2004.837669]

[39]   H. Chen, Y. Wang, and R. Zhou, "Transient and voltage stability enhancement via coordinated excitation and UPFC control", *Generation, Transmission and Distribution, IEE Proceedings,* vol. 148, 2001pp. 201-208

[40]   M. Januszewski, J. Machowski, and J.W. Bialek, "Application of the direct Lyapunov method to improve damping of power swings by control of UPFC", *Generation, Transmission and Distribution, IEE Proceedings,* vol. 151, 2004pp. 252-260

[http://dx.doi.org/10.1049/ip-gtd:20040054]

[41] G. Andersson, P. Donalek, R. Farmer, N. Hatziargyriou, I. Kamwa, and P. Kundur, "Causes of the 2003 major grid blackouts in North America and Europe, and recommended means to improve system dynamic performance", *Power Systems, IEEE Transactions on,* vol. 20, pp. 1922-1928, 2005.
[http://dx.doi.org/10.1109/TPWRS.2005.857942]

[42] M.D. Ilic, H. Allen, W. Chapman, C.A. King, J.H. Lang, and E. Litvinov, "Preventing Future Blackouts by Means of Enhanced Electric Power Systems Control: From Complexity to Order", *Proc. IEEE,* vol. 93, pp. 1920-1941, 2005.
[http://dx.doi.org/10.1109/JPROC.2005.857496]

[43] J. Machowski, J.W. Bialek, and J.R. Bumby, Power System Dynamics and Stability, ed: John Wiley and Sons.
[http://dx.doi.org/10.1016/0140-6701(96)88716-9]

[44] J.D. Glover, M. Sarma, and T. Overbye, *Power Systems Analysis and Design.* Cengage Learning, 2007.

[45] K.R. Padiyar, *Power System Dynamics.* BS Publiations: New Delhi, 2008.

[46] A.F. Abdou, A. Abu-Siada, and H.R. Pota, "Damping of subsynchronous oscillations and improve transient stability for wind farms", In: *Innovative Smart Grid Technologies Asia (ISGT), 2011 IEEE PES,* 2011, pp. 1-6.
[http://dx.doi.org/10.1109/ISGT-Asia.2011.6167077]

[47] R.G. Farmer, A.L. Schwalb, and E. Katz, "Navajo project report on subsynchronous resonance analysis and solutions", *Power Apparatus and Systems, IEEE Transactions on,* vol. 96, pp. 1226-1232, 1977.
[http://dx.doi.org/10.1109/T-PAS.1977.32445]

[48] W. Li, and L. Ching-Huei, "Application of dynamic resistance braking on stabilizing torsional oscillations", *TENCON '93. Proceedings. Computer, Communication, Control and Power Engineering.1993 IEEE Region 10 Conference on,* vol. 5, 1993pp. 145-148

[49] "First benchmark model for computer simulation of subsynchronous resonance", *Power Apparatus and Systems, IEEE Transactions on,* vol. 96, pp. 1565-1572, 1977.
[http://dx.doi.org/10.1109/T-PAS.1977.32485]

[50] L. Ha Thu, and S. Santoso, "Increasing wind farm transient stability by dynamic reactive compensation: Synchronous-machine-based ESS versus SVC", *Power and Energy Society General Meeting, 2010 IEEE,* 2010pp. 1-8
[http://dx.doi.org/10.1109/PES.2010.5589756]

[51] H. Zhou, H. Wei, X. Qiu, J. Xu, X. Wei, and S. Wang, "Improvement of Transient Voltage Stability of the Wind Farm Using SVC and TCSC", *Power and Energy Engineering Conference (APPEEC),* 2011pp. 1-4
[http://dx.doi.org/10.1109/APPEEC.2011.5749161]

[52] A.F. Abdou, A. Abu-Siada, and H.R. Pota, "Application of SVC on stabilizing torsional oscillations and improving transient stability", *Power and Energy Society General Meeting, 2012 IEEE,,* 2012pp. 1-5
[http://dx.doi.org/10.1109/PESGM.2012.6344674]

[53] A. Abu-Siada, "Damping of large turbo-generator subsynchronous resonance using superconducting magnetic energy storage unit", *Universities Power Engineering Conference (AUPEC),* 2010pp. 1-4

[54] J.B. Devotta, M.G. Rabbani, and S. Elangovan, "Application of superconducting magnetic energy storage unit for damping of subsynchronous oscillations in power systems", *Energy Convers. Manage.,* vol. 40, pp. 23-37, 1999.
[http://dx.doi.org/10.1016/S0196-8904(98)00108-3]

[55] W. Li, L. Shin-Muh, and H. Ching-Lien, "Damping subsynchronous resonance using superconducting

magnetic energy storage unit", *Energy Conversion, IEEE Transactions on,* vol. 9, pp. 770-777, 1994.
[http://dx.doi.org/10.1109/60.368329]

[56]   O. Wasynczuk, "Damping Subsynchronous Resonance Using Energy Storage", *Power Engineering Review, IEEE,* vol. PER-2, pp. 36-37, 1982.
[http://dx.doi.org/10.1109/MPER.1982.5519415]

[57]   A.F. Abdou, A. Abu-Siada, and H.R. Pota, "Application of a STATCOM for damping subsynchronous oscillations and transient stability improvement", *Universities Power Engineering Conference (AUPEC),* 2011pp. 1-5

[58]   E. Eitelberg, J.C. Balda, E.S. Boje, and R.G. Harley, "Stabilizing SSR oscillations with a shunt reactor controller for uncertain levels of series compensation", *Power Systems, IEEE Transactions on,* vol. 3, pp. 936-943, 1988.
[http://dx.doi.org/10.1109/59.14544]

[59]   L. Wang, and C-H. Lee, "Stabilizing torsional oscillations using a shunt reactor controller", *Energy Conversion, IEEE Transactions on,* vol. 6, pp. 373-380, 1991.
[http://dx.doi.org/10.1109/60.84309]

[60]   R.M. Hamouda, Z.R. AlZaid, and M.A. Mostafa, "Damping torsional oscillations in large turbo-generators using Thyristor Controlled Braking Resistors", *Power Engineering Conference,* 2008pp. 1-6

[61]   A. Ghorbani, and S. Pourmohammad, "A novel excitation controller to damp subsynchronous oscillations", *Int. J. Electr. Power Energy Syst.,* vol. 33, pp. 411-419, 2011.
[http://dx.doi.org/10.1016/j.ijepes.2010.10.002]

[62]   W. Li, "Damping of torsional oscillations using excitation control of synchronous generator: the IEEE Second Benchmark Model investigation", *Energy Conversion, IEEE Transactions on,* vol. 6, pp. 47-54, 1991.
[http://dx.doi.org/10.1109/60.73788]

[63]   F.D. Jesus, E.H. Watanabe, L.F. Souza, and J.E. Alves, "Analysis of SSR Mitigation Using Gate Controlled Series Capacitors", *Power Electronics Specialists Conference,* 2005pp. 1402-1407
[http://dx.doi.org/10.1109/PESC.2005.1581813]

[64]   R. Rajaraman, I. Dobson, R.H. Lasseter, and S. Yihchih, "Computing the damping of subsynchronous oscillations due to a thyristor controlled series capacitor", *Power Delivery, IEEE Transactions on,* vol. 11, pp. 1120-1127, 1996.
[http://dx.doi.org/10.1109/61.489376]

[65]   A. Abu-Siada, and S. Islam, "Application of SMES Unit in Improving the Performance of an ac/dc Power System", *IEEE Transactions on Sustainable Energy,* vol. 2, pp. 109-121, 2011.
[http://dx.doi.org/10.1109/TSTE.2010.2089995]

[66]   G. Abad, J. López, M.A. Rodríguez, L. Marroyo, and G. Iwanski, "Introduction to A Wind Energy Generation System", In: *Doubly Fed Induction Machine, ed: John Wiley and Sons, Inc.,*, 2011, pp. 1-85.
[http://dx.doi.org/10.1002/9781118104965.ch1]

[67]   A.F. Abdou, H.R. Pota, A. Abu-Siada, and Y.M. Alharbi, "Application of STATCOM-HTS to improve DFIG performance and FRT during IGBT short circuit", *Power Engineering Conference (AUPEC),* 2014pp. 1-5
[http://dx.doi.org/10.1109/AUPEC.2014.6966641]

[68]   A.M. Shiddiq-Yunus, A.A. Siada, and M.A. Masoum, "Impact of SMES on DFIG Power Dispatch during Intermittent Misfire and Fire-through Faults", *IEEE Trans. Appl. Supercond.,* 2012.

[69]   J. Arrillaga, *High Voltage Direct Current Transmission.* Peter Peregrinus Ltd: London, 1983.

[70]   J. Arrillaga, *High Voltage Direct Current Transmission.* Institution of Electrical Engineers, 1998.
[http://dx.doi.org/10.1049/PBPO029E]

[71]   G. Abad, J. López, M.A. Rodríguez, L. Marroyo, and G. Iwanski, "Back-to-Back Power Electronic Converter", In: *Doubly Fed Induction Machine, ed: John Wiley and Sons, Inc.,*, 2011, pp. 87-154. [http://dx.doi.org/10.1002/9781118104965.ch2]

[72]   H.B. Sethom, and M.A. Ghedamsi, "Intermittent Misfiring Default Detection and Localisation on a PWM Inverter Using Wavelet Decomposition", *Journal of Electrical System,* vol. 4, pp. 222-234, 2008.

[73]   K.R. Padiyar, *HVDC Power Transmission Systems.* John Wiley & Sons: New Delhi, 1990.

[74]   T. Ackermann, *Wind Power in Power System.* John Wiley and Sons Ltd: West Sussex, 2005. [http://dx.doi.org/10.1002/0470012684]

[75]   L. Bin, and S. Sharma, "A survey of IGBT fault diagnostic methods for three-phase power inverters", *International Conference on Condition Monitoring and Diagnosis,* 2008pp. 756-763 [http://dx.doi.org/10.1109/CMD.2008.4580396]

[76]   M. Mohseni, and S.M. Islam, "Review of international grid codes for wind power integration: Diversity, technology and a case for global standard", *Renew. Sustain. Energy Rev.,* vol. 16, pp. 3876-3890, 2012. [http://dx.doi.org/10.1016/j.rser.2012.03.039]

[77]   V. Akhmatov, Induction Generators for Wind Power: Multi-Science Pub., 2005.

CHAPTER 3

Superconducting Magnetic Energy Storage, a Promising FACTS Device for Wind Energy Conversion Systems

Abstract: The applications of FACTS devices have become popular in the last few decades. There are many types of FACTS devices that are currently used in power systems to improve system stability, power quality and the overall reliability of the power systems. Since the involvement of renewable energies based power plants such as wind and PV, problems related to power system stability and quality has become even more complex, therefore the deployment of FACTS devices has become a challenging task. In this chapter, a Superconducting Magnetic Energy Storage (SMES) Unit is applied to improve the performance of Doubly Fed Induction Generator (DFIG) based wind turbine during various disturbances such as voltage sag, short circuit faults and load variation, including problems related to internal faults within the DFIG converters.

Keywords: Superconducting Magnetic Energy Storage, Wind Energy conversion system, Fault ride through, Active and reactive power modulation.

INTRODUCTION

The first generation of wind energy conversion systems (WECS) was the direct connected WECS type. This technology uses a fixed speed turbine to generate power and it dominated renewable energy installations worldwide, comprising up to 70% of all installations in 1995 [1]. This technology remained popular until the electronic power revolution that updated the WECS so that they could maximize wind energy capture. This technology is based on variable speed WECS, and it can optimally capture wind energy 5% more than the fixed speed WECS option. Furthermore, the variable speed WECS can reduce the impact of transient wind gusts and subsequent fatigue unlike the fixed speed turbines [2, 3]. Due to the extensive penetration of WECS into the existing power networks, the transmission system operators (TSO) have developed new requirements for integrating WECS to the grid, which are known as grid codes [4]. One of the important parts of the grid codes is the voltage profile at the point of common coupling (PCC), which determines the connection or disconnection status of WECS from the grid. Two

Ahmed Abu-Siada, Mohammad A.S. Masoum, Yasser Alharbi, Farhad Shahnia & A.M. Shiddiq Yunus

strategies can improve the performance or the fault ride through (FRT) capability of the WECS. Most literatures have studied the path of developing new control techniques to fulfil the criterion of TSO [5 - 9]. However, this strategy is effective only for new installations and new connections of WECS to the grid. Alternatively, designers can apply a flexible ac transmission system (FACTS) device or storage energy system [10], which is more cost effective for large existing WECS.

Variable speed WECS, such as doubly fed induction generator (DFIG) and Type D WECS (or so called full converter WECS), were introduced to overcome the limitations of the fixed speed type in capturing maximum wind energy and to contribute reactive power to the grid when required [11]. Compared to full scale variable speed WECS, DFIG is very sensitive to grid faults [6]; although the DFIGs are usually connected far away from the grid, the grid faults affect the voltage profile at the PCC. The significant voltage drop/rise at the PCC causes high currents in the stator windings and may induce over-current in the rotor windings. Moreover, during a grid fault, voltage drops at the DFIG terminal, high current flows at both grid and rotor side converters and high voltage at the dc link may lead to converter station blocking. This condition will force the DFIG to be disconnected from the system. If the DFIG contributes a large portion of power to the grid, then the financial loss of this disconnection would be uncountable. Most of the studies on DFIG address the improvement of its FRT capability during voltage sag [5 - 9]. No attention, however, has been given to improving the DFIG performance under voltage sag and voltage swell conditions using the same controller. Both sag and swell can lead to the disconnection of WECS when the voltage profile during disturbances occurring at the grid side violates the minimum and maximum voltage levels set by the grid code.

Numerous studies have explored the ability of the superconducting magnetic energy storage (SMES) to smooth out WECS power output. Most of these studies have connected SMES to a fixed speed WECS [12 - 18]. These studies, however, could not claim that SMES are more effective than BESS or hybrid energy storages in term of cost and capability. The only situations that would make better use of SMES than other storage options are the multi-purpose conditions, such as improving the FRT capability of WECS during sag and swell events in the grid side, load levelling and during converter faults event.

As a part of this chapter, a short load variation (load levelling) on a power system with DFIG is also studied. During this load variation, the SMES unit is able to compensate the variation close to the predetermined power that DFIG must transfer. Moreover, a new study of misfire and fire-through that may take place within both the grid-side converter (GSC) and the rotor-side converter (RSC) of the DFIG is also included to compare the performance of using UPFC (chapter 2) and the SMES unit. The impacts of these switching faults on the performance of DFIG with and without SMES are investigated and discussed.

SYSTEM UNDER STUDY

The system under study is shown in Fig. (**3.1**). It consists of six-1.5 MW DFIG connected to the ac grid at PCC via Y/Δ step up transformer. The grid is represented by an ideal 3-phase voltage source of constant frequency and is connected to the wind turbines via 30 km transmission line. The reactive power produced by the wind turbine is regulated at 0 MVAr at normal operating conditions. For an average wind speed of 15 m/s which is used in this study, the turbine output power is 1.0 pu and the generator speed is 1.0 pu. SMES Unit is connected to the 25 kV bus and is assumed to be fully charged at its maximum capacity of 1.0 MJ.

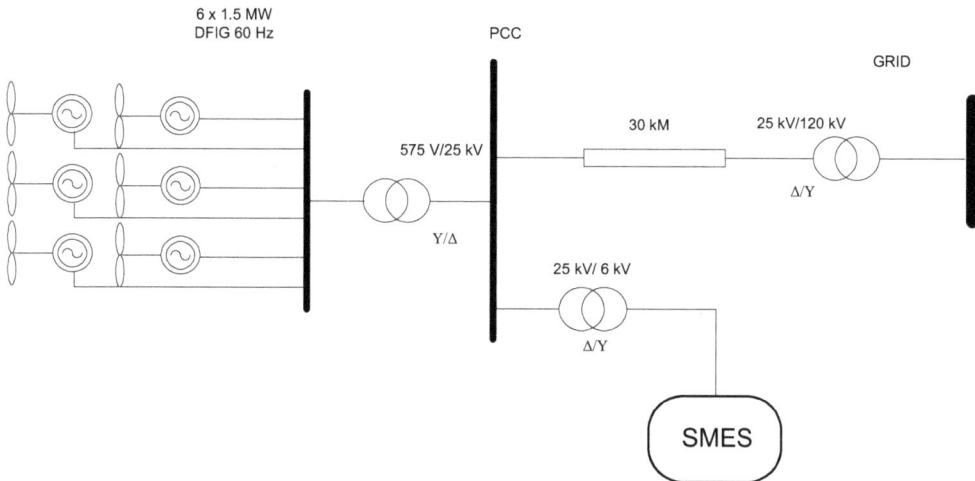

Fig. (3.1). System under study.

SMES UNIT CONFIGURATION

Superconducting Magnetic Energy Storage (SMES) is a storage system that operates in superconductivity state. The SMES magnetic coil is placed in a very low temperature called "Cryogenic Temperature" typically between 32°K to127°K depending on the coil material to maintain the resistance value of the coil close to zero. This condition makes the ability of the coil to store energy optimally because of its very low losses. Moreover, since the SMES system has no moving part such as in pumped hydro and flywheel energy storage, 90 to 99% efficiency of SMES unit can be achieved. The stored energy in the SMES coil can be calculated from:

$$E_{SM} = \frac{1}{2} L_{SM} I_{SM}^2 \qquad (3.1)$$

where E_{SM} is the SMES energy in Joule; I_{SM} is the SMES Current (A) and L_{SM} is the SMES coil inductance.

The SMES unit configuration consists of VSC and dc-dc Chopper which are connected through a dc capacitor link. The VSC is controlled by a hysteresis current controller (HCC) while the dc-dc chopper is controlled by fuzzy logic controller (FLC) as configured out in Fig. (**3.2**).

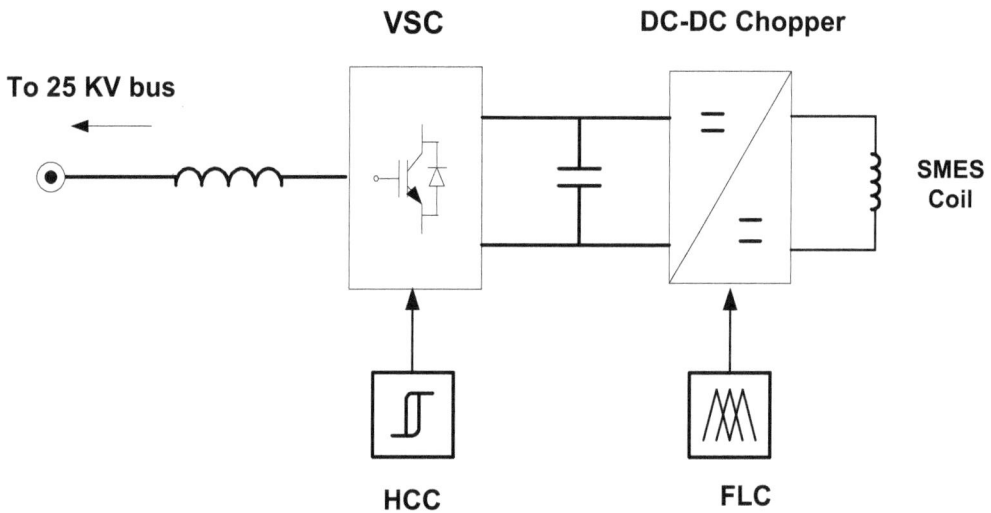

Fig. (3.2). SMES configuration.

Among the various PWM technique, the hysteresis band current control is used very often because of its simplicity of implementation, fast response current loop, and the method does not need any knowledge of load parameters. To allow charging and discharging process of SMES coil energy, dc-dc chopper is applied with the SMES coil. The control system of the dc-dc chopper is operated by fuzzy logic control (FLC). The capacitor is used to store a constant dc voltage that is maintained by VSC controller. The active power from the generator and the current in the superconductor coil are used as inputs to the fuzzy logic controller. To determine the value of duty cycle of the dc chopper, active power of the DFIG and SMES coil current are used as inputs of the fuzzy logic controller. The duty cycle is compared with 1000 Hz saw-tooth signal to produce signal for the dc-dc Chopper as explained in detail in the following session.

SMES CONTROLLER

The proposed control algorithm for the SMES unit is much simpler and closer to realistic applications compared with the SMES controller proposed in Refs [14, 20]. In these publications, the four proposed PI controllers require more computational time to optimally tune their parameters to maintain the overall system stability and to achieve satisfactory dynamic response during transient events. Moreover, the control system for the dc-dc chopper in these studies has only considered the generated active power (P_G) as a control parameter but did not take into account the energy capacity of the SMES unit. The proposed control scheme here comprises two PI controllers and considers the SMES coil current to take the SMES stored energy capacity into account along with the DFIG generated power as control parameters to determine the direction and level of power exchange between the SMES coil and the ac system. This control system is efficient, simple and easy to implement as will be elaborated below.

The proposed SMES with an auxiliary phase locked loop (PLL) controller is shown in Fig. (**3.3**) The hysteresis current controller (HCC) is comparing the 3-phase line currents (I_{ABC}) with the reference currents ($I_{ABC}*$) which is dictated by the I_D* and I_Q* references. The values of I_D* and I_Q* are generated through the conventional PI controllers based on the error values of V_{dc} and V_S. The value of I_D* and I_Q* is converted through the Park's transformation (*dq0-abc*) to produce the reference current ($I_{ABC}*$).

To control the power transfer between SMES coil and the ac system, a dc-dc chopper is used and fuzzy logic is selected to control its duty cycle (*D*) as shown in Fig. (**3.4**). The fuzzy logic controller (FLC) is developed according to the fuzzy inference which is a process of formulating the mapping from a given input to the designated output. Input variables for the model are the real power generated by the DFIG and SMES coil current (Fig. **3.5**).

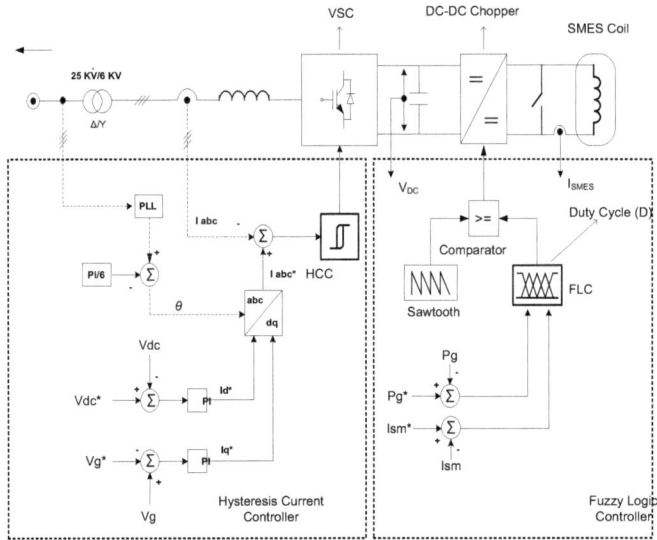

Fig. (3.3). SMES unit configuration and the proposed HCC-FLC control scheme [19].

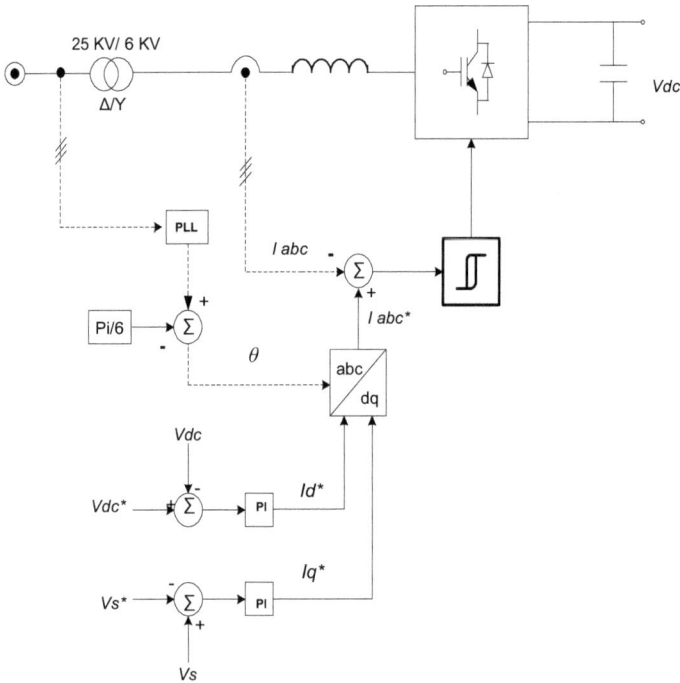

Fig. (3.4). Control algorithm of VSC.

The output of FLC is D for class D dc-dc chopper that is shown in Fig. (**3.6**). The V-I operational range for the SMES coil is shown in Fig. (**3.7**). The duty cycle determines the direction and magnitude of power exchange between the SMES coil and the ac system as presented in Table **3.1**.

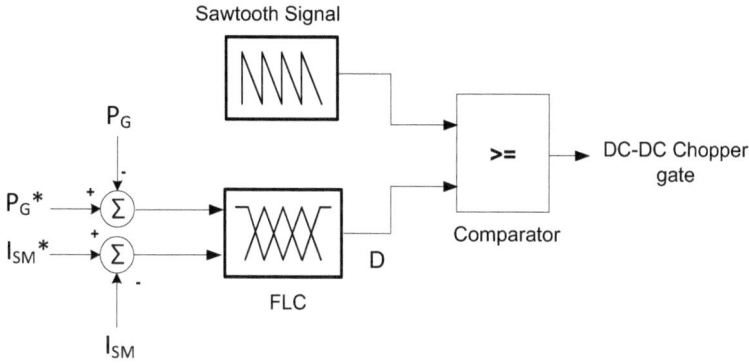

Fig. (3.5). Control algorithm of dc-dc chopper.

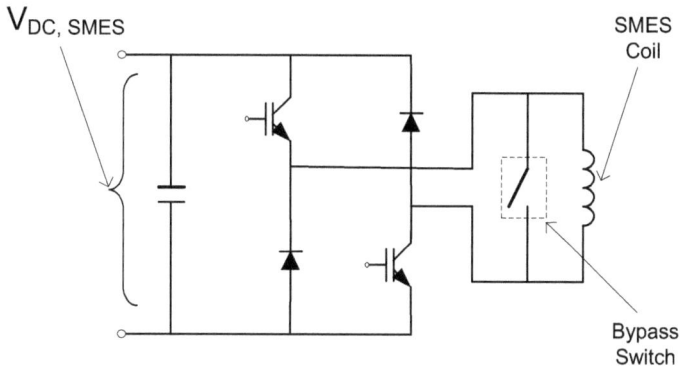

Fig. (3.6). Class D dc-dc chopper topology with a SMES coil.

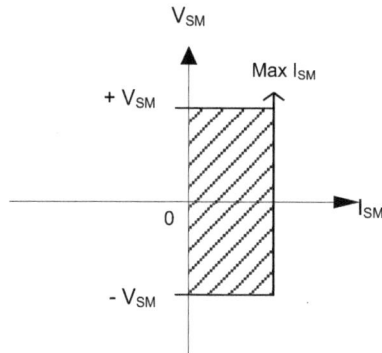

Fig. (3.7). Operation range of SMES coil.

Table 3.1. Rules of duty cycle

Duty Cycle (D)	SMES Coil Action
D = 0.5	Standby Condition
$0 \leq D < 0.5$	Discharging Condition
$0.5 < D \leq 1$	Charging Condition

If the duty cycle D is equal to 0.5, no action will be taken by the coil and the system is under normal operating condition. In this condition, a bypass switch that is installed across the SMES coil (shown in Fig. **3.6**) will be closed to avoid the draining process of SMES energy during normal condition. The bypass switch is controlled in such a way it will be closed if D is equal to 0.5 otherwise it will be opened. This technique has been introduced in some studies in the literature [14, 20]. When the grid power is reduced, D will be reduced accordingly to be in the range of 0 to 0.5 and the stored energy in the SMES coil will be transferred to the ac system. Charging process of the SMES coil takes place when D is in the range of 0.5 to 1.0.

The relation between V_{SMES} and $V_{dc, SMES}$ can be written as [21]:

$$V_{SMES} = (1 - 2D)V_{DC,SMES} \tag{3.1}$$

where:

V_{SMES} is the average voltage across the SMES coil (V)

D is the duty cycle

$V_{dc,SMES}$ is the average voltage across the dc link capacitor (V)

The proposed fuzzy model is built using the graphical user interface tool provided by MATLAB. Each input was fuzzified into five sets of gaussmf type membership function (MF). The Gaussian curve is a function of a vector, x, and depends on parameters σ and c as given by:

$$f(x,\sigma,c) = e^{-(x-c)^2/2\sigma^2} \tag{3.2}$$

where:

σ is a variable that determines the center of the peak
c is the width of the bell curve

The corresponding membership functions for the input variables P_G and I_{SMES} are shown in Fig. (**3.8** and **3.9**), respectively. The membership functions for the output variable (duty cycle) are considered on the scale 0 to 1.0 as shown in Fig. (**3.10**).

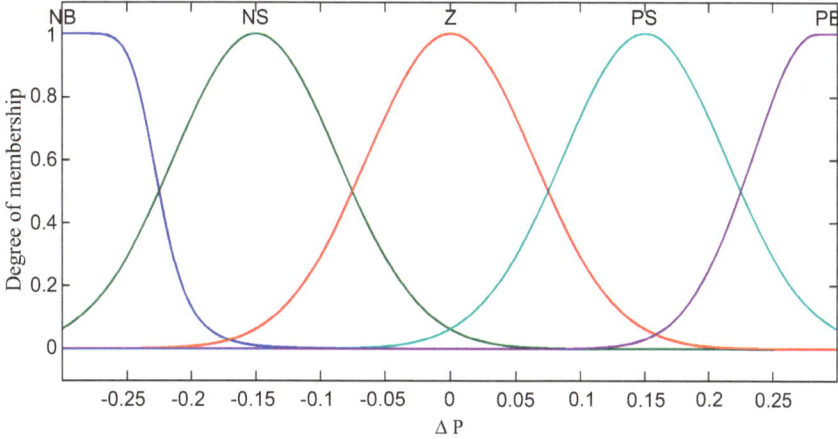

NB=Negative Big, NS= Negative Small, Z= Zero, PS=Positive Small, PB=Positive Big

Fig. (3.8). Memberships function for the input variable P_G (pu).

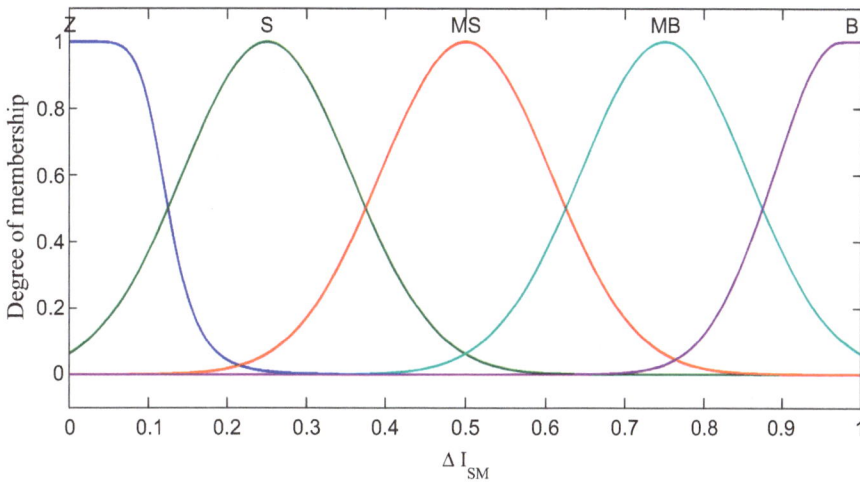

Z=Zero, S= Small, MS=Medium Small, MB= Medium Big, B= Big

Fig. (3.9). Memberships function for the input variable I_{SMES} (pu).

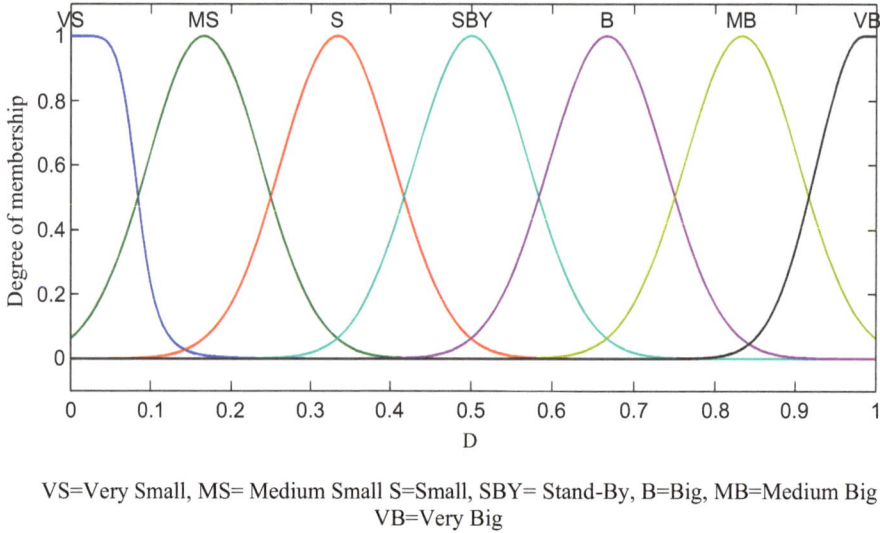

VS=Very Small, MS= Medium Small S=Small, SBY= Stand-By, B=Big, MB=Medium Big
VB=Very Big

Fig. (3.10). Memberships function for the output variable *D* (duty cycle).

Centre-of-gravity which is widely used in fuzzy models is used for defuzzification process where the desired output z_0 is calculated as [22]:

$$z_0 = \frac{\int z.\mu_c(z)dz}{\int \mu_c(z)dz} \tag{3.3}$$

where:

$\mu_c(z)$ is the membership function of the output.

The variation range in SMES current and DFIG output power, as well as the corresponding duty cycle are used to develop a set of fuzzy logic rules in the form of (IF-AND-THEN) statements to relate the input variables to the output. The duty cycle for any set of input variable (P_G and I_{SMES}) can be evaluated from the surface graph shown in Fig. (**3.11**).

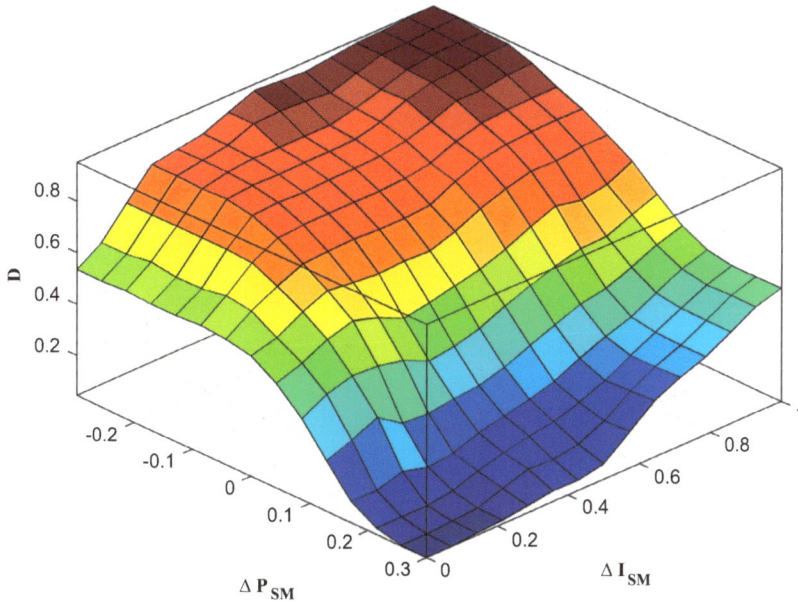

Fig. (3.11). Surface graph-duty cycle.

APPLICATION OF SMES ON DFIG-BASED WECS DURING VOLTAGE SAG

Power quality issue is the common consideration for new construction or connection of power generation system including WECS installation and their connection to the existing power system. Voltage sag (dip) and swell at the grid side are investigated to assess the compliance of the fault ride through capability of WECS equipped with DFIG. Voltage sag is a decrease of the rms voltage to a level between 0.1 and 0.9 pu at the power frequency for a duration of 0.5 cycles to 1.0 minute. Voltage sags are usually associated with system faults but can also be caused by switching of heavy loads or starting of large motors [23, 24].

A voltage sag depth of 0.5 pu lasting for 0.05 s is applied at t=2.0 s at the grid side of the system under study. Without the SMES unit, the real power produced by the DFIG drops to 0.6 pu and it reaches maximum overshooting of 40% during the clearance of the fault as shown in Fig. (**3.12a**). As can be seen in Fig. (**3.12a**), with the SMES unit connected to the system, DFIG output power will drop to only 0.875 pu. Fig. (**3.12b**) implies that with the connection of the SMES unit and during the event of voltage sag the reactive power support by the DFIG is reduced and the steady state condition is reached faster compared with the system without SMES. The voltage at the PCC is shown in Fig. (**3.12c**), where without SMES, voltage drops to 0.6 pu. However, by connecting the SMES unit, voltage drop at the PCC is reduced to only 0.8 pu which leads to a voltage drop at the generator terminal of 0.8 pu that is referenced as a safety margin by the wind turbine manufacturers [25, 26].

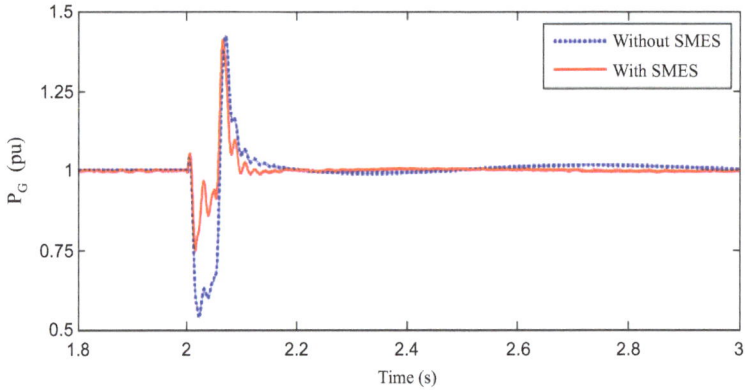

(a)

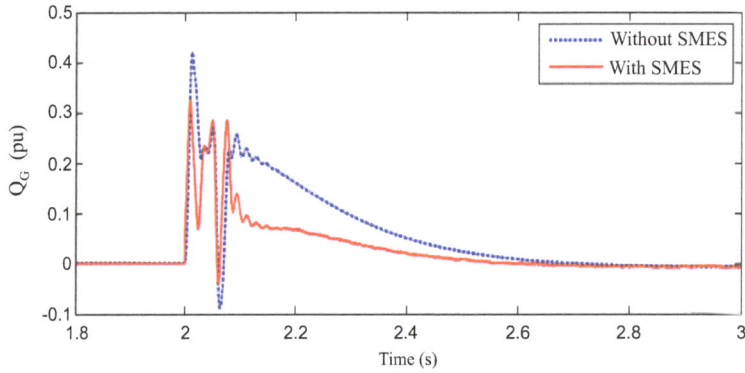

(b)

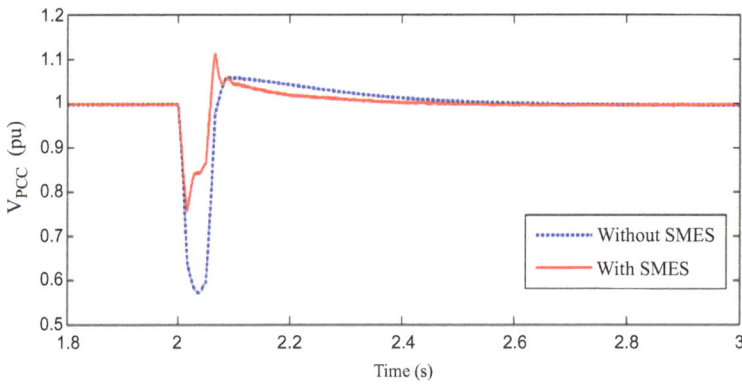

(c)

Fig. 3.12 contd.....

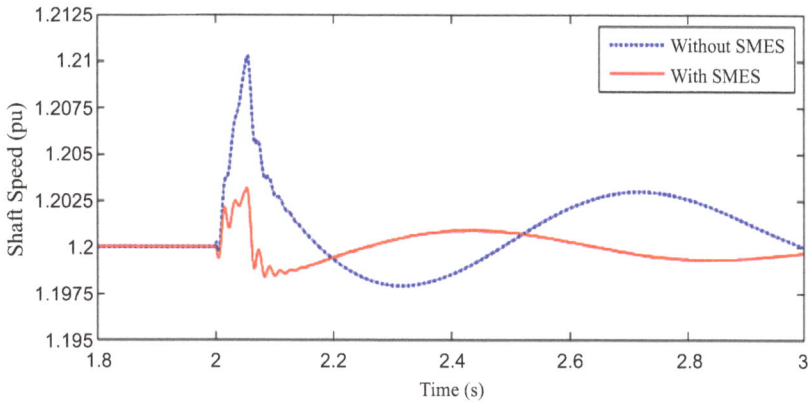

(d)

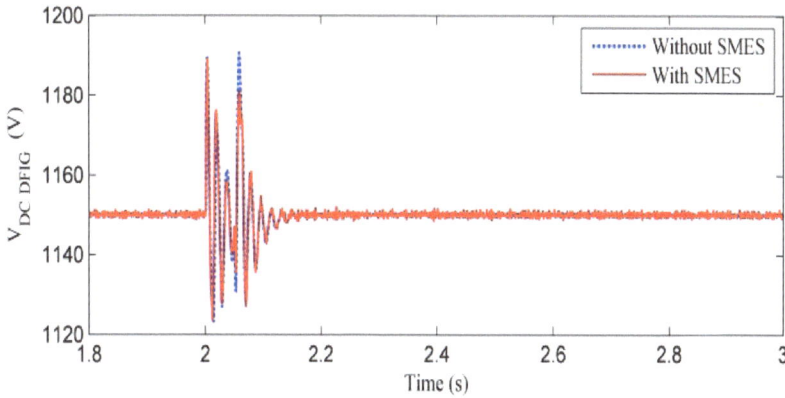

(e)

Fig. (3.12). DFIG responses during voltage sag without/with a SMES unit; (a) Active power, (b) Reactive power, (c) PCC voltage, (d) Shaft speed, and (e) Voltage at dc link of DFIG.

The DFIG power drop causes the generator speed to be accelerated to compensate for the power imbalance. As can be observed in Fig. (**3.12d**), the generator speed accelerates and oscillates without the SMES unit; however, with the SMES connected to the system, the power drop is reduced, the settling time of the generator speed is substantially reduced and the overshooting level is significantly decreased. Another effect of the voltage sag on the DFIG's behavior is on the voltage across the DFIG dc link capacitor that is shown in Fig. (**3.12e**). The voltage overshoot across the DC link capacitor during fault clearance is slightly reduced with the SMES unit connected to the system.

APPLICATION OF SMES ON DFIG-BASED WECS DURING VOLTAGE SWELL

A swell is defined as an increase in the rms voltage at the power frequency for durations from 0.5 cycles to 1.0 minute. Typical magnitudes are between 1.1 and 1.8 pu. As with dips, swells are usually associated with system fault conditions, but they are much less common than voltage dips. A swell can occur due to a single line-to-ground fault on the system resulting in a temporary voltage rise on the un-faulted phases. Swells can also be caused by switching off a large load or switching on a large capacitor bank [23, 24].

In this simulation, a voltage swell is applied by increasing the voltage level at the grid side to 1.5 pu. The voltage swell is assumed to start at t= 2.0 s and lasts for 0.05 s. In this event, DFIG generated power will increase upon the swell occurrence and will be reduced when it is cleared as shown in Fig. (**3.13a**).

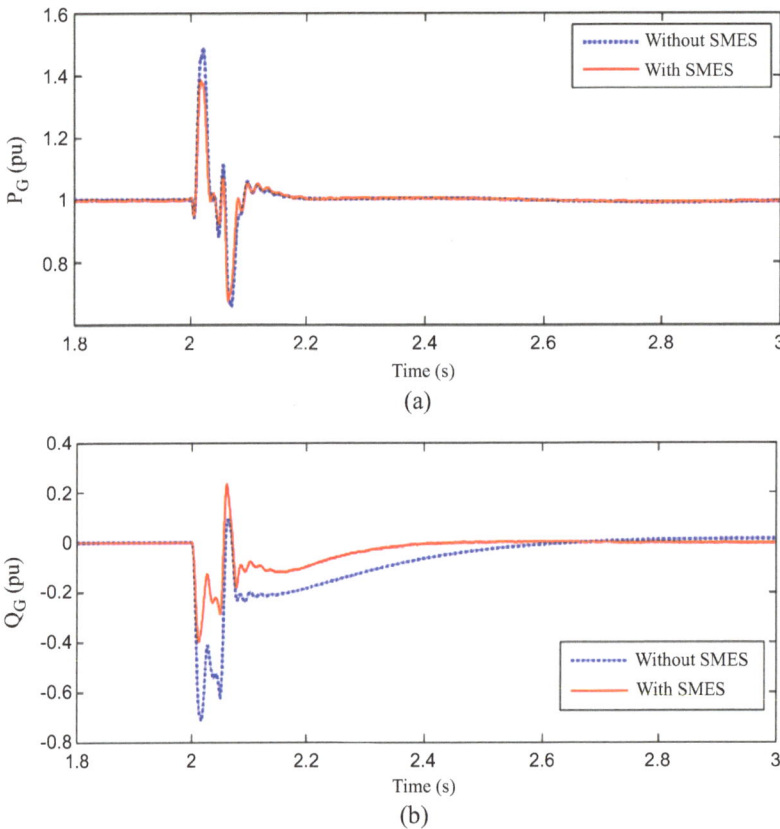

(a)

(b)

Fig. 3.13 contd.....

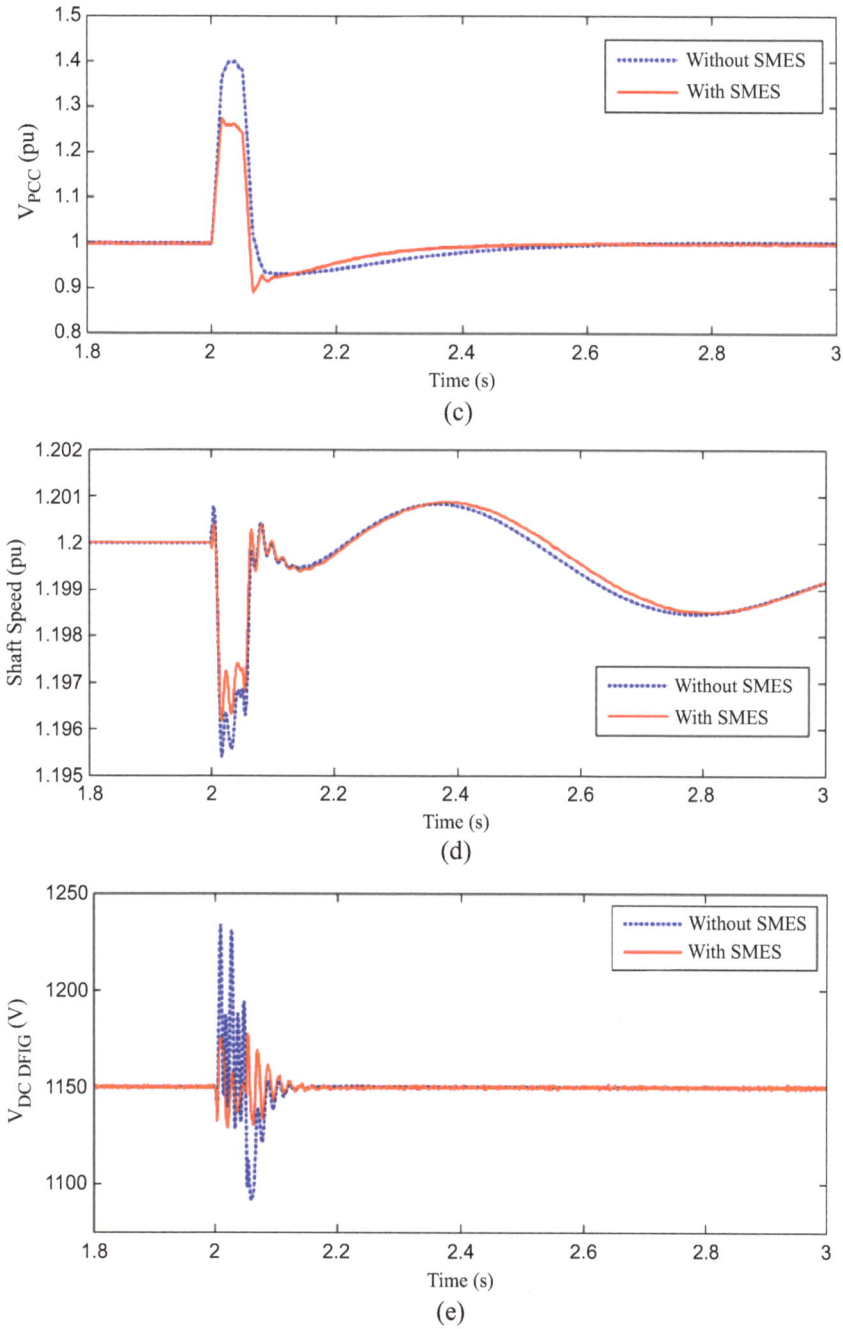

Fig. (3.13). DFIG responses during voltage swell without/with a SMES unit; (a) Active power, (b) Reactive power (c) PCC voltage (d) Shaft speed, and (e) Voltage at dc link of DFIG.

The maximum power overshoot is slightly reduced with the SMES unit connected to the system. To compensate for the voltage rise, DFIG absorbs the surplus reactive power as shown in Fig. (**3.13b**). The amount of reactive power absorbed by the DFIG is lesser with SMES connected to the PCC since the voltage profile at the PCC is rectified to a level below 1.3 pu with the connection of the SMES unit while this voltage remains above 1.3 pu without SMES connected to the PCC (Fig. **3.13c**). Without the connection of the SMES unit, the voltage at the PCC does not comply with the HVRT of Spain and Australia grid codes [4, 27] which will lead to the disconnection of the DFIG from the system. The shaft speed (shown in Fig. (**3.13d**) is slightly improved with the connection of SMES unit to the system. Without the connection of the SMES unit, the voltage across the DFIG dc link capacitor will experience significant oscillations and overshooting level upon voltage swell incidence as can be shown in Fig. (**3.13e**). In some occasions, this may lead to the blocking of the converters [25]. As shown in Fig. (**3.13e**), voltage oscillations as well as voltage overshooting level are significantly reduced by connecting SMES unit to the system.

APPLICATION OF SMES ON DFIG SYSTEM DURING SMALL DISTURBANCE AND LOAD VARIATION

The load frequency control issue occurs as a result of sudden and small load perturbation which creates an instantaneous mismatch between real power supply and demand. This problem can be controlled by the governor action in conventional thermal power plants. The load frequency control problem is attributed to the fact that the inertia of the rotating parts is the only storage capacity in a conventional power system where the additional power demand can be met through the kinetic energy of the generator rotor. This will limit the degree to which frequency variation can be minimized by appropriate governor control. The problem is more serious in case of WECS based on DFIG in which the inertia is much lower than their power rating. Solving this problem by attaching a flywheel to the rotor shaft to increase its inertia will significantly increase the torsional stress on the shaft during dynamic oscillation. Since the wind turbine generated power depends on wind speed which cannot be controlled, the wind turbine can only be down-regulated to match the power demand by applying additional controllers [27]. However, when loads are increased above the rated power output of the generated power, in particular when wind speed is low, power imbalance takes place causing the load frequency control problem. The case of such situation had occurred for example, on the European outage on November, 2006, caused the tripping of 4892 MW of wind-origin power in Western Europe exacerbating the imbalance between demand and supply in this area [28]. Adding a SMES unit which has a very fast time response to the load bus can improve the system overall performance during such conditions [29].

The system under study is slightly modified from the previous study as shown in Fig. (**3.14**). For study of the dynamic system, a local load is added at the PCC. Fig. (**3.15**) shows the load profile used in the simulation studies. It is assumed that the nominal load power is constant under normal operating condition and is fed through the power generated from the wind turbine. The load is assumed to experience a ± 0.2 pu fluctuation for a duration of 0.5s at t= 3.0s and t= 4.0s respectively.

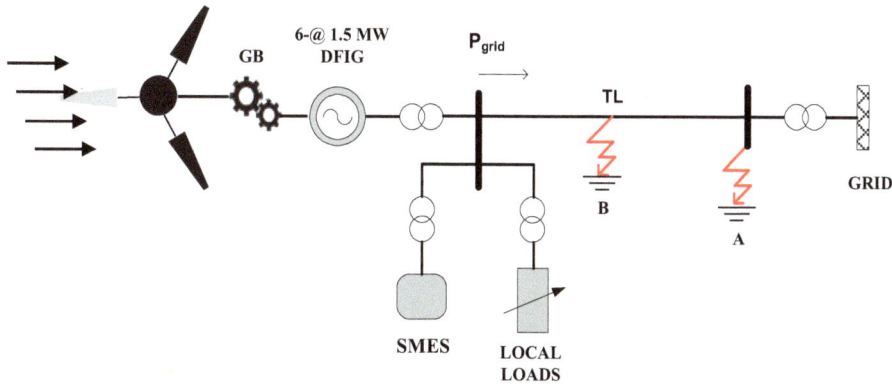

Fig. (3.14). Investigated system for load variation effect [30].

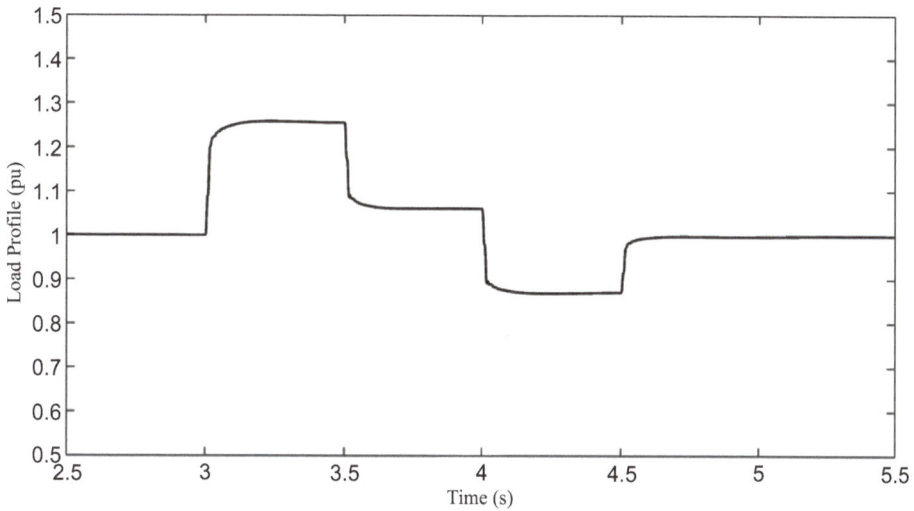

Fig. (3.15). Load profile under study.

In this short load variation study, the control rules of the fuzzy set are modified as provided in Figs. (**3.16** to **3.19**).

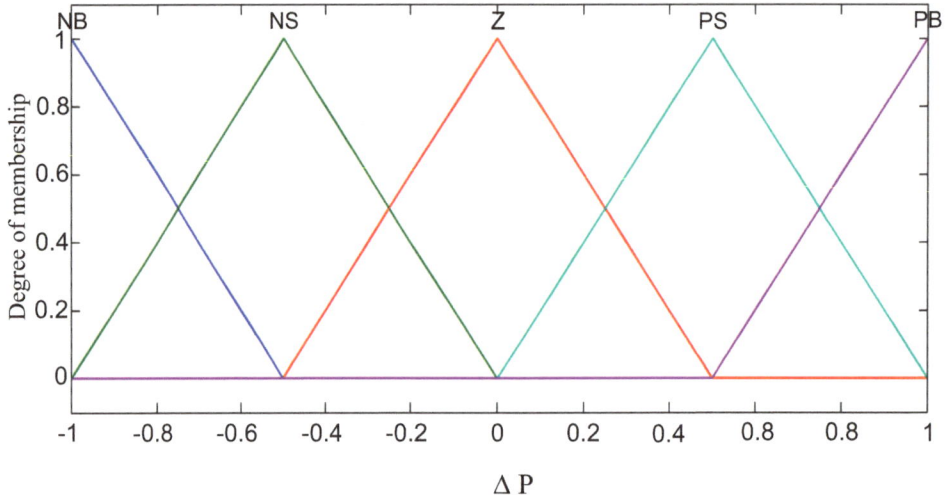

NB=Negative Big, NS=Negative Small, Z=Zero, PS=Positive Small, PB=Positive Big

Fig. (3.16). Memberships function for the input variable P_G (pu).

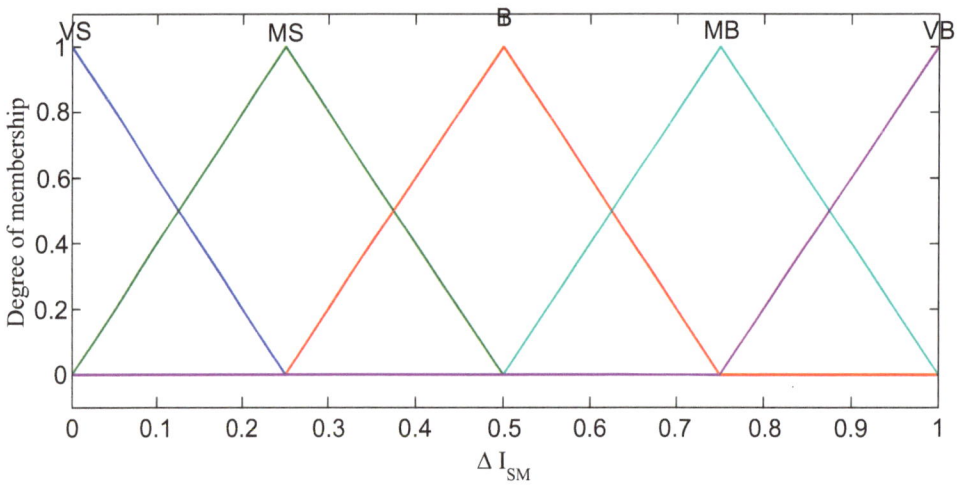

VS=Very Small, MS=Medium Small, B=Big, MB=Medium Big, VB=Very Big

Fig. (3.17). Memberships function for the input variable I_{SMES} (pu).

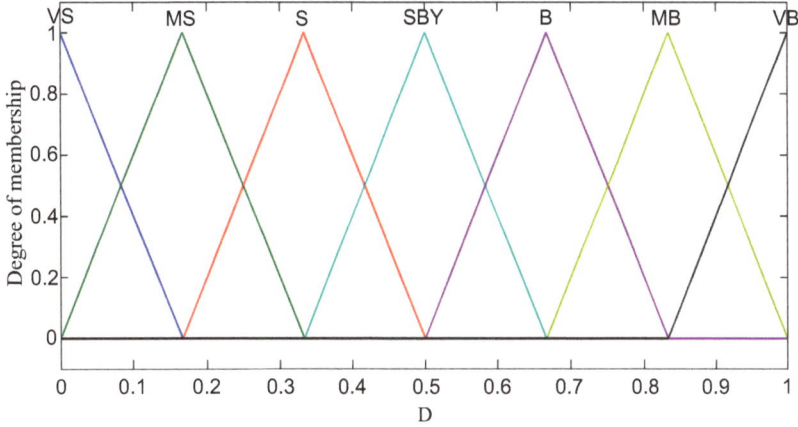

VS=Very Small, MS=Medium Small, S=Small, SBY=Stand-By, B=Big, MB=Medium Big, VB=Very Big

Fig. (3.18). Memberships function for the output variable D (duty cycle).

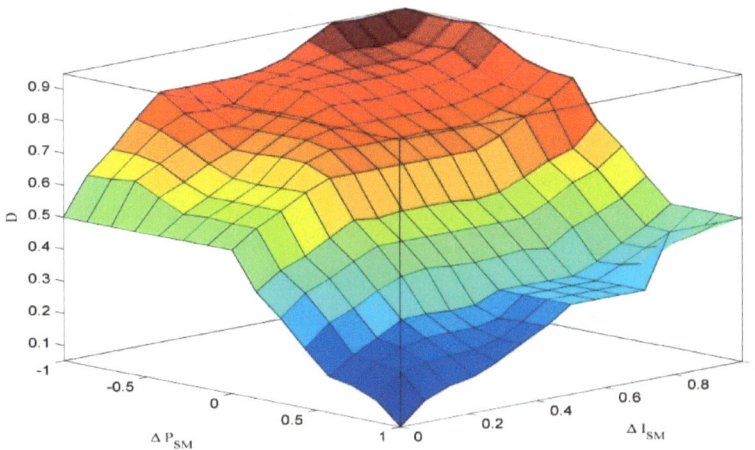

Fig. (3.19). Surface graph-duty cycle.

Fig. (**3.20a**) shows the power transferred to the grid through the transmission line. When load power demand increases at t= 3.0 pu, the power transfer through the transmission line is decreasing by 30% and it increases by only 10% due to transmission power transfer capability when the load is decreased at t= 4.0s. By connecting the SMES unit to the DFIG bus, the two levels can be substantially decreased to only 10% and 5% respectively. The dc-dc chopper duty cycle D is shown in Fig. (**3.20b**) where at normal operating conditions, the value of D is maintained by the FLC at a level of 0.5 and the SMES coil current is held constant at its rated value, consequently, there will be no energy transfer between SMES unit and the grid as can be shown from the coil stored energy in Fig. (**3.20c**). When the load increases suddenly at t= 3.0s, the controller will act to decrease the value of D below 0.5 level and energy stored in the coil will be discharged to the grid to provide active power support. When the load returns to its steady state level, the coil will be partially charged but not to its full capacity. When the load is decreased at t= 4.0s, the value of D increases due to the controller action to a level above 0.5 and the surplus power in the system will be used to fully charge the SMES coil.

LARGE DISTURBANCE

To examine the ability of the SMES unit to improve the DFIG dynamic performance during a large disturbance, a six-cycle three-phase short circuit fault is applied at t= 0.5s and is assumed to last for 6 cycles. Two fault locations are considered; case A where the short circuit fault is assumed to take place at the grid terminals (point A in Fig. **3.14**) and case B where the fault is located at the middle of the transmission line (point B in Fig. **3.14**).

Case A: Fig. (**3.21a**) shows the active power generated by the DFIG is reduced by 75% at the instant of fault occurrence and it experiences a maximum overshooting of 60% at the instant of fault clearance. As a result of generated power reduction, the shaft speed accelerates during the fault and experiences some oscillations after fault clearance as shown in Fig. (**3.21b**). The large amount of power loss will definitely influence the supply for local loads and the power transfer to the grid. If the DFIG is the main power supply to the local load, the economic loss will be significant. The voltage across the DFIG dc-link capacitor (shown in Fig. (**3.21c**) experiences significant oscillations and overshooting level through fault duration; however, its maximum overshooting remains within the safety margin of 1.2 pu as specified in [25]. The voltage at the PCC is reduced to a level of 0.4 p.u. during the fault. When compared to the Spain fault ride through requirement, the voltage profile at the PCC violates the permissible lower limit as shown in Fig. (**3.21c**) which calls for the disconnection of WECS from the grid. When the short circuit is cleared, the DFIG converters act to provide reactive power support to the grid

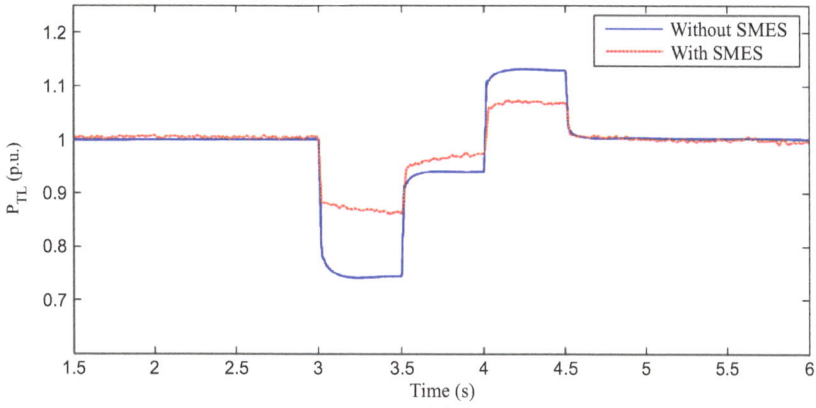

(a)

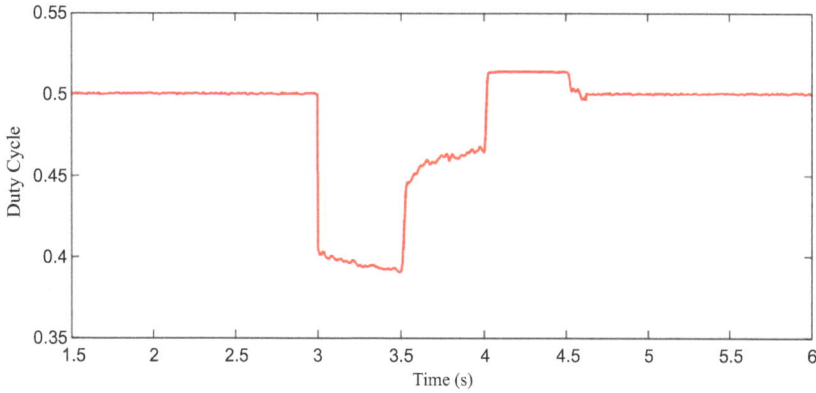

(b)

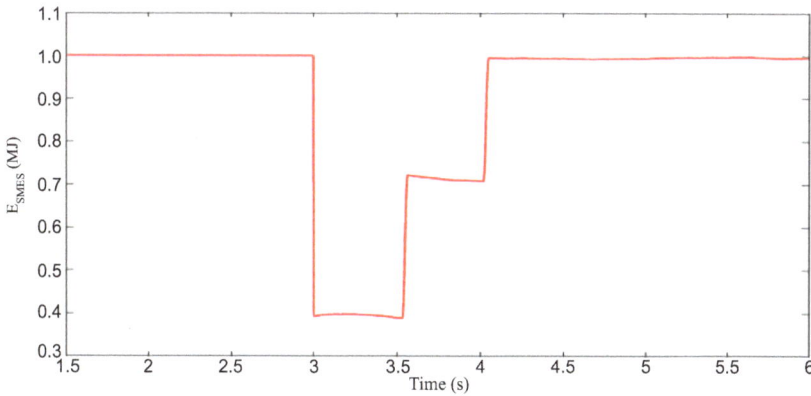

(c)

Fig. (3.20). (a) Power transfer to the grid and (b) Duty cycle response during dynamic event (c) Stored energy of SMES during dynamic event.

to recover the voltage to its rated value [31]. By connecting the SMES unit to the DFIG bus, the proposed controller acts to change the duty cycle level and hence the energy transfer direction according to the system requirement. As shown in Fig. (**3.21e**), the duty cycle is reduced below 0.5 during the fault and the entire SMES stored energy is discharged to support the system during the fault as shown in Fig. (**3.21f**). As a result, the DFIG generated power will be reduced by only 30% and its maximum overshooting and settling time are substantially reduced as shown in Fig. (**3.21a**). As a result of this improvement, the acceleration and settling time of the rotor shaft speed are significantly reduced as can be seen in Fig. (**3.21b**). There is a slight improvement in the DFIG dc-link voltage (Fig. **3.20c**) in terms of the over shooting levels but as mentioned above, these levels are within the safety margins specified in [25]. Connecting the SMES unit to the system will bring the voltage at the PCC to the safety margin of Spain grid codes as shown in Fig. (**3.21d**). In this case, the WECS connection to the grid can be maintained to support the grid during the fault duration.

Case B: When 3-phase short circuit fault occurs at the middle of the transmission line, DFIG generated power drops significantly as shown in Fig. (**3.22a**). The impact of this fault on the shaft speed and dc- link capacitor voltage is worse when compared to the previous case as shown in Figs. (**3.22b and c**), respectively. Voltage level at the PCC reaches 0.1 p.u. and it violates the low voltage ride through level of Spain grid code as shown in Fig. (**3.22d**). With the SMES unit connected to the PCC bus; generated power drop can be improved compared to the system without SMES unit as shown in Fig. (**3.22a**). Also, shaft speed and dc link voltage shown in Figs. (**3.22b and c**) are improved with the connection of the SMES unit. With the connection of the SMES unit, the connection of the wind turbine to the grid during such fault can be maintained as shown in Fig. (**3.22d**), where the voltage drop at the PCC is brought to a safety margin of the Spain's FRT. The trend of duty cycle and energy stored within the superconductor coil is similar to case A however; they are slightly larger than those in case A as shown in Figs. (**3.22e and f**).

APPLICATION OF SMES ON WECS DURING CONVERTER FAULTS

Impact of SMES on DFIG Power Dispatch During Intermittent Misfire and Fire-Through Faults

Misfire and Fire-Through

In DFIG-based WECS, the DFIG is interfaced to the ac network through GSC and RSC which are considered as the crux of the wind power system. The converter stations determine the ability of wind turbine to optimally operate during wind

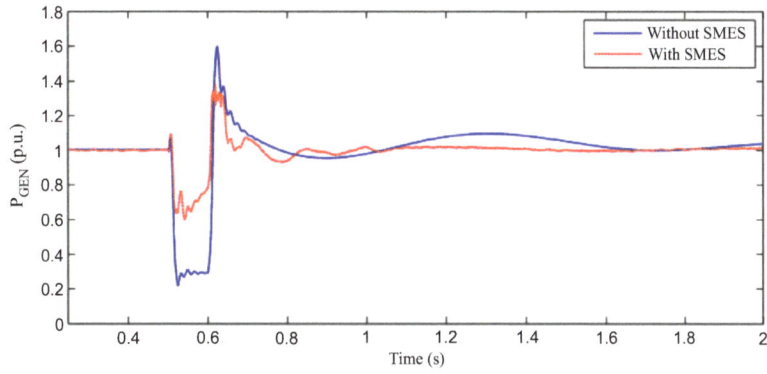

(a)

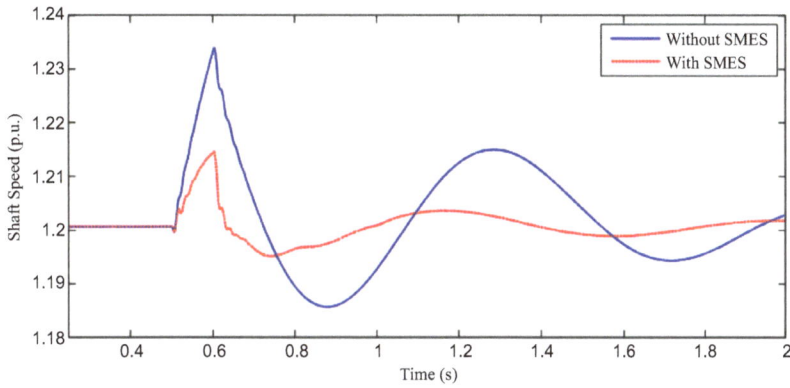

(b)

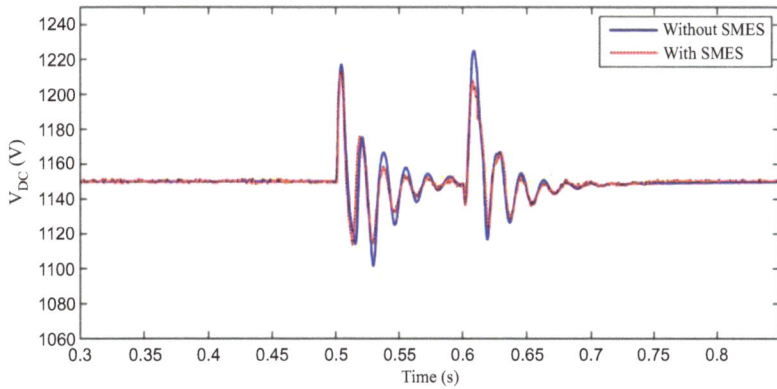

(c)

Fig. 3.21 contd.....

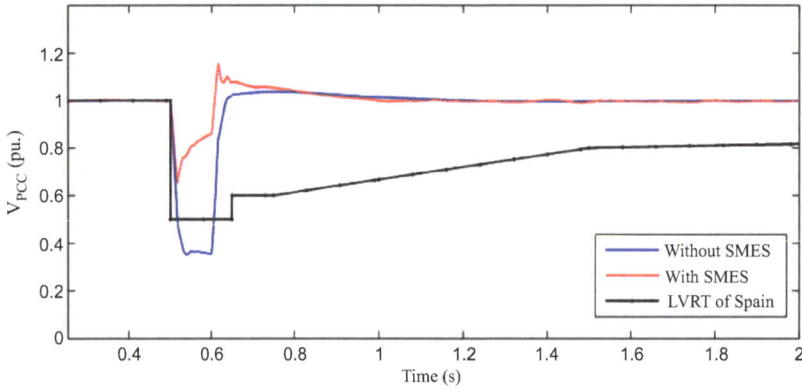

(d)

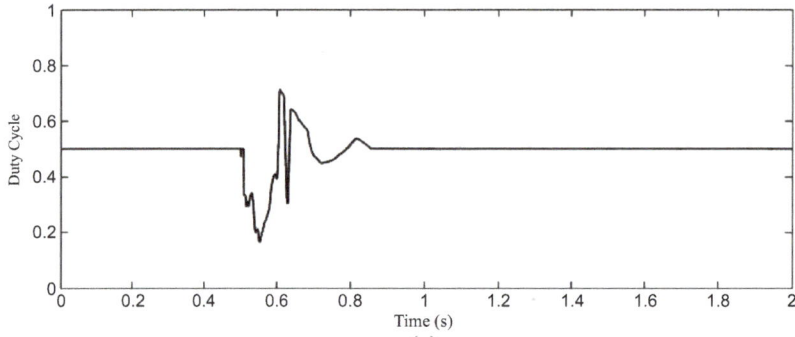

(e)

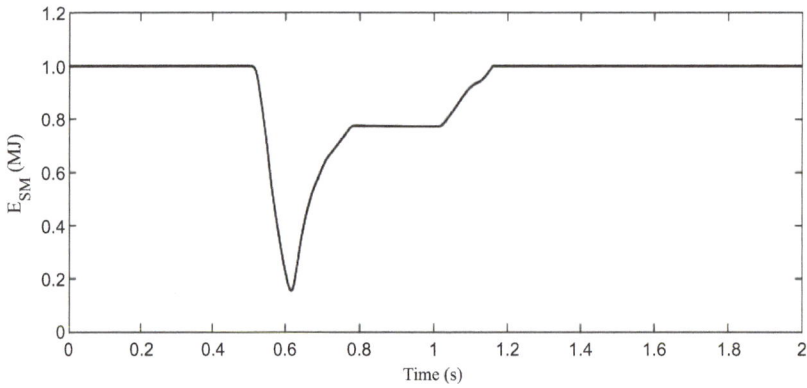

(f)

Fig. (3.21). Response of the DFIG during 3-phase short circuit fault at point A; (a) Generated power of DFIG, (b) Rotor shaft speed and (c) Voltage across DFIG dc-link (d) Voltage profile at PCC, (e) Duty cycle response of the SMES unit, and (f) Stored energy response of the SMES unit

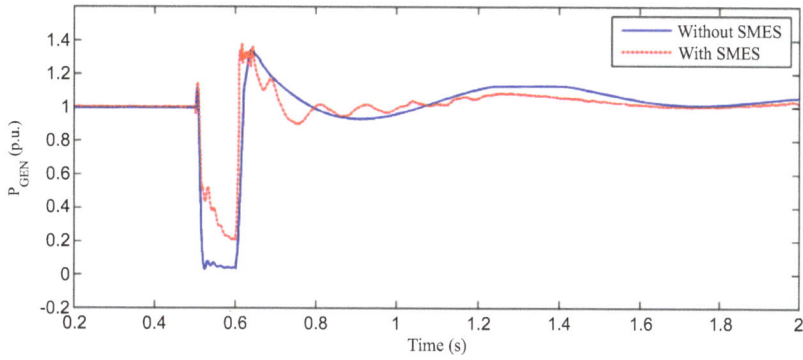

(a)

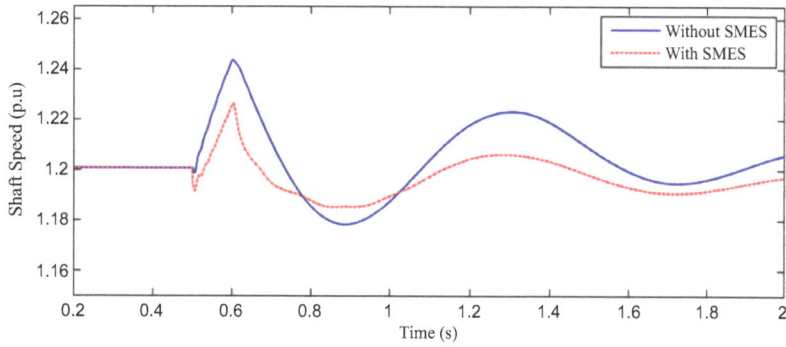

(b)

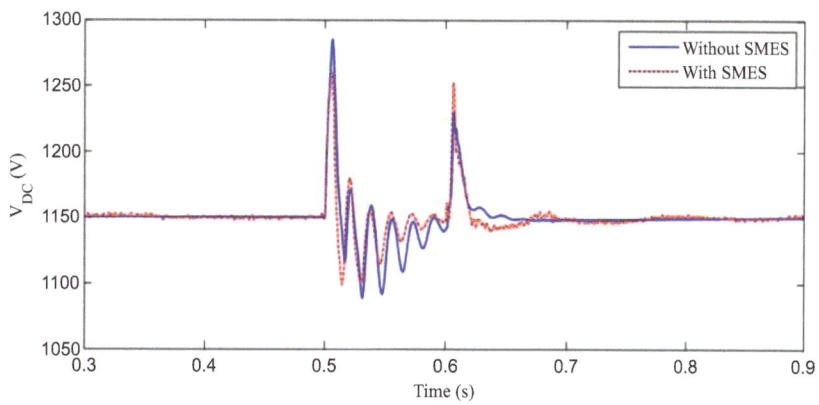

(c)

Fig. 3.22 contd.....

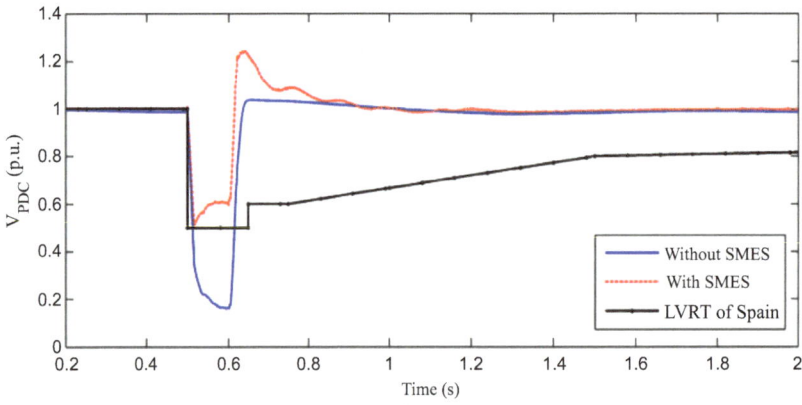

(d)

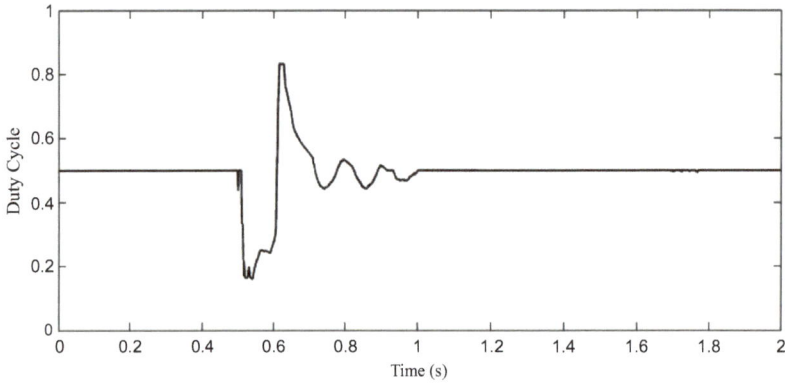

(e)

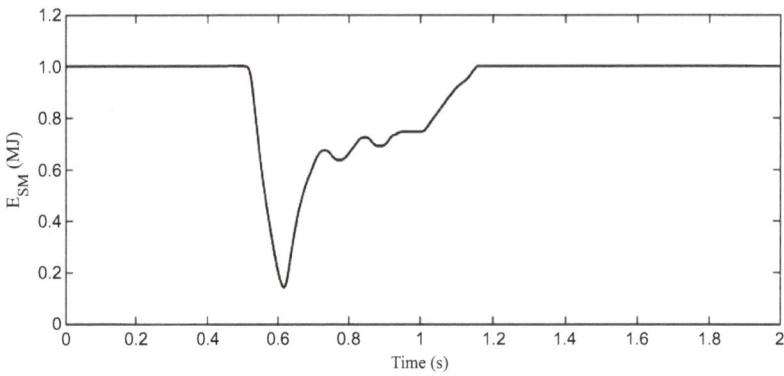

(f)

Fig. (3.22). Response of the DFIG during 3-phase short circuit fault at point B; (a) Generated power of DFIG, (b) Rotor shaft speed, (c) Voltage across DFIG dc-link, (d) Voltage profile at PCC, (e) Duty cycle response of the SMES unit and (f) Stored energy response of the SMES unit.

speed fluctuation and can provide reactive power support during grid disturbance events. Therefore, any faults within the converter stations will affect the overall performance of the DFIG. Due to the utilization of two converters, it is very vulnerable for the DFIG to have switching malfunctions on its converters such as misfire and fire-through. Misfire is defined as the failure to fire a valve during a scheduled conducting period and fire-through is the failure to block a valve during a scheduled non-conducting period. These faults are caused by various malfunctions in the control and firing equipment [32]. Some of these converter faults are self-clearing if the causes that led to these faults are of transient nature. However they can still cause a major problem to the system particularly when they occur on inverter station instead of the rectifier station [33]. When gate-misfiring occurs on IGBT based converter, it can cause catastrophic breakdown to the device, if the faults remains undetected [34]. Although switching of renewable energy can lead to power quality issue [24], with proper small step-wise control of the ac transmission line power, this frequency variation can be controlled better as discussed in Ref [35]. However, as mentioned in [32], the switching malfunctions are still possible to occur. To clearly understand the system under study, the DFIG along with the SMES is re-configured as shown in Fig. (**3.23**). The control algorithm concept and all parameters of the SMES unit are the same as the case of grid sag and swell discussed above. Most of the studies in the literatures dedicated to the DFIG performance during various grid disturbances such as voltage sag [36 - 41]. Although, there are few studies in the literatures about the effect of converter faults on the overall performance of HVDC system [42 - 44] and misfiring detection on PWM inverter [45], there is no attention given to investigate the performance of the DFIG during misfire and fire-through faults. In this case study, both misfire and fire-through are applied to the IGBT-1 in the RSC and GSC of Fig. (**3.24**).

The simulation of misfire and fire-through in DFIG's converters is demonstrated in Figs. (**3.25** and **3.26**) respectively. The faults are assumed to start at 0.1 s and lasts for 10 ms as shown in the IGBT-1 pulse signals for GSC and RSC of the DFIG. In the case of misfire the generated signals from PWM for both GSC and RSC are failed to fire the IGBT-1 valve in its scheduled conducting period as shown in Figs. (**3.25a and b**). For fire-through case, shown in Figs. (**3.26a and b**), the generated switching signals from PWM are failed to block its valve during a scheduled non-conducting period.

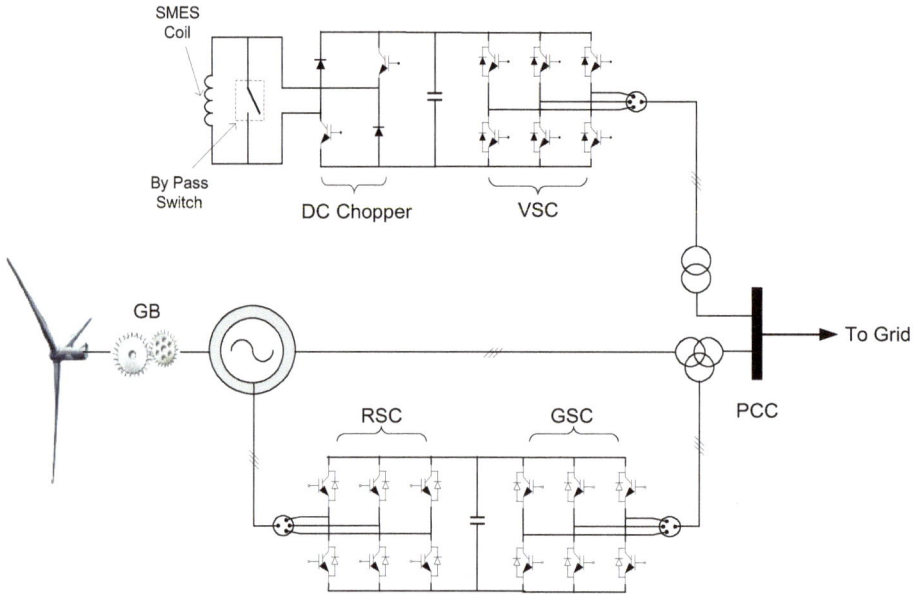

Fig. (3.23). System under study for DFIG-SMES with misfire and fire-through [46].

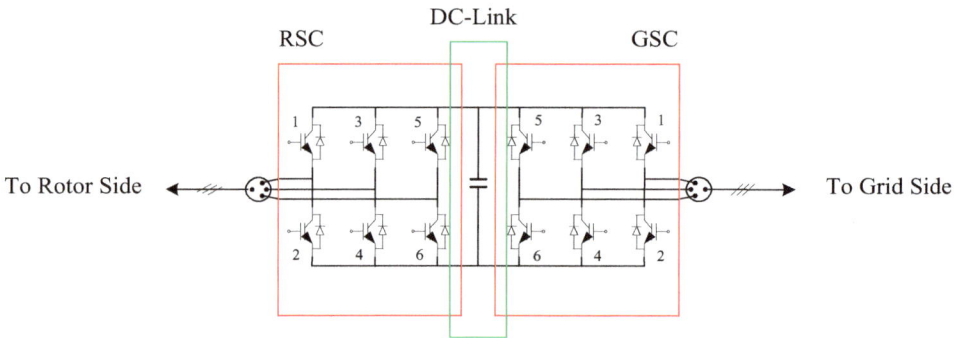

Fig. (3.24). Converters configuration of the DFIG.

Misfire on GSC and RSC

In this study, intermittent misfire is simulated in both GSC and RSC of the DFIG based WECS. In all studied cases, the fault is assumed to take place on switch-1 (as shown in Fig. **3.24**) at t= 0.50s and self-healed at t= 0.55s.

When a misfire is applied to the GSC, the generated power (P), shaft speed and the voltage at PCC (V_{PCC}) are not significantly influenced, this is attributed to the fact that GSC has no direct connection with the DFIG's rotor and hence its influence on its dynamic performance is trivial. This is evidenced by the slight oscillations that the above parameters experiencing during fault period as shown in Fig. (**3.27**).

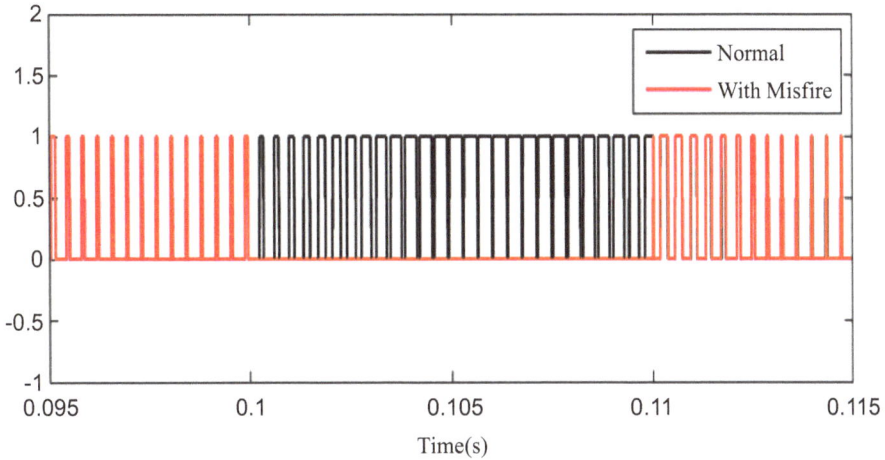

(a)

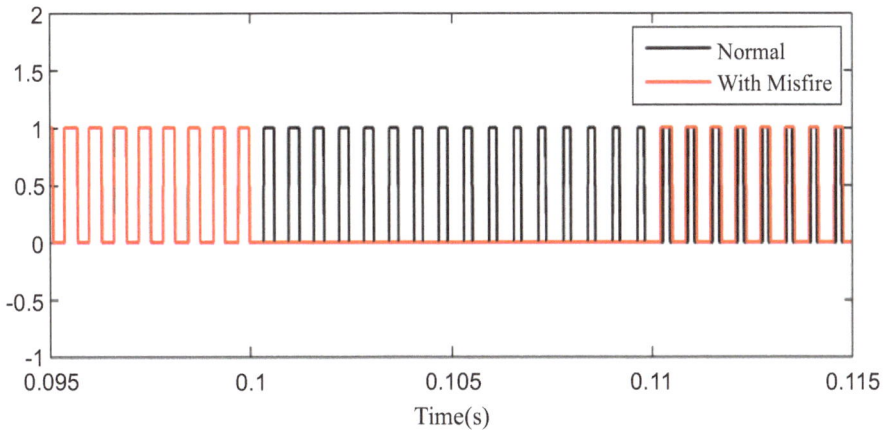

(b)

Fig. (3.25). Simulation of misfire fault in IGBT-1; (a) GSC signal and (b) RSC signal.

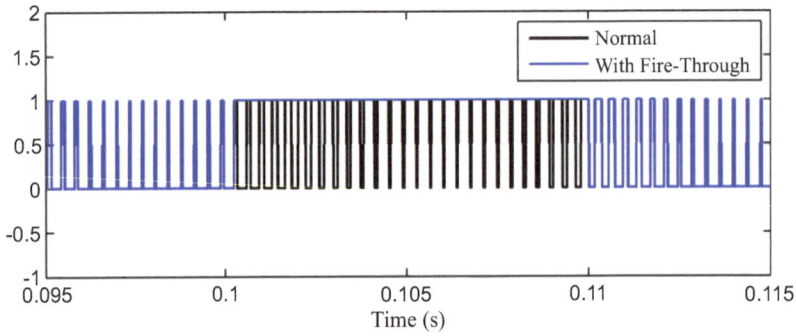

(a)

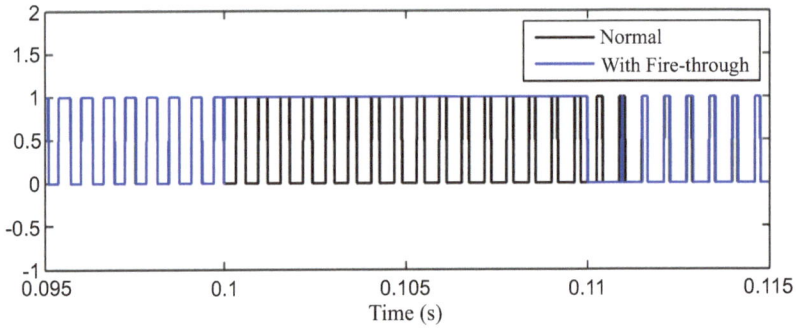

(b)

Fig. (3.26). Simulation of Fire-through fault in IGBT-1; (a) GSC signal and (b) RSC signal.

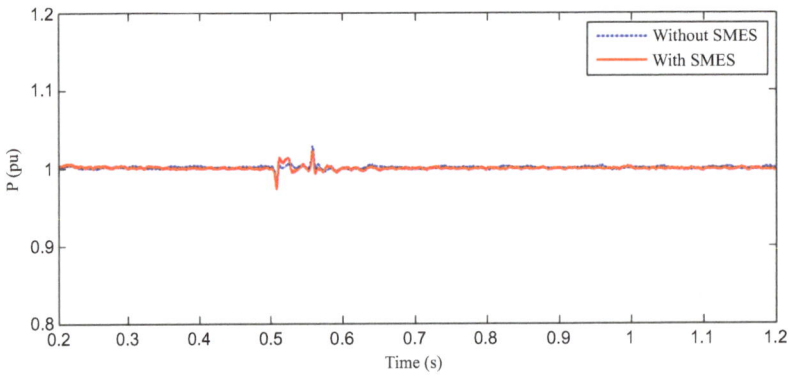

(a)

Fig. 3.27 contd.....

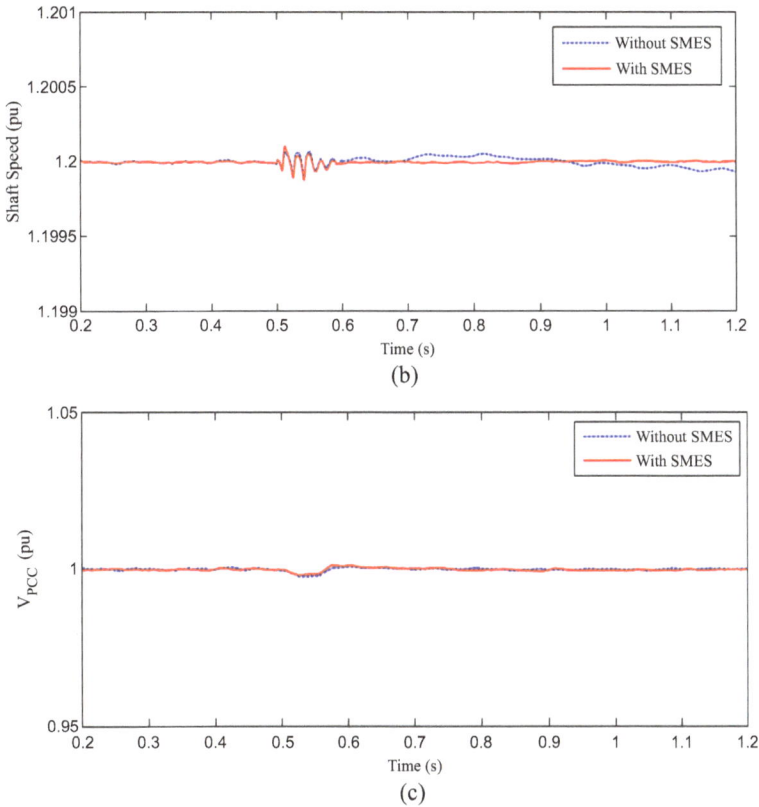

Fig. (3.27). Effect of GSC misfireon DFIG dynamic responses with and without SMES coils; (a) Power and (b) Shaft speed (c) Voltage at PCC.

The SMES unit when connected to the system slightly reduces the oscillations and the settling time however, its contribution is not significant as all variables are within safe margins.

When misfire is applied to RSC, the DFIG generated power is reduced dramatically by 60% as shown in Fig. (**3.28a**), shaft speed exhibits maximum overshooting at the instant of fault occurrance and it does not settle down to the origional level of 1.2 pu as shown in Fig. (**3.28b**), and the voltage at the PCC is reduced by 6% (Fig. **3.28c**). SMES unit can modulate both active and reactive power and support the system during the fault. Thus by connecting SMES unit to the system, the generated power reduction will be only 20% as shown in Fig. (**3.26a**). The overshooting in shaft speed is reduced and the settling time is substantially decreased as shown in Fig. (**3.28b**). The voltage at the PCC is significantly improved during the fault and after fault clearance as can be shown in Fig. (**3.28c**).

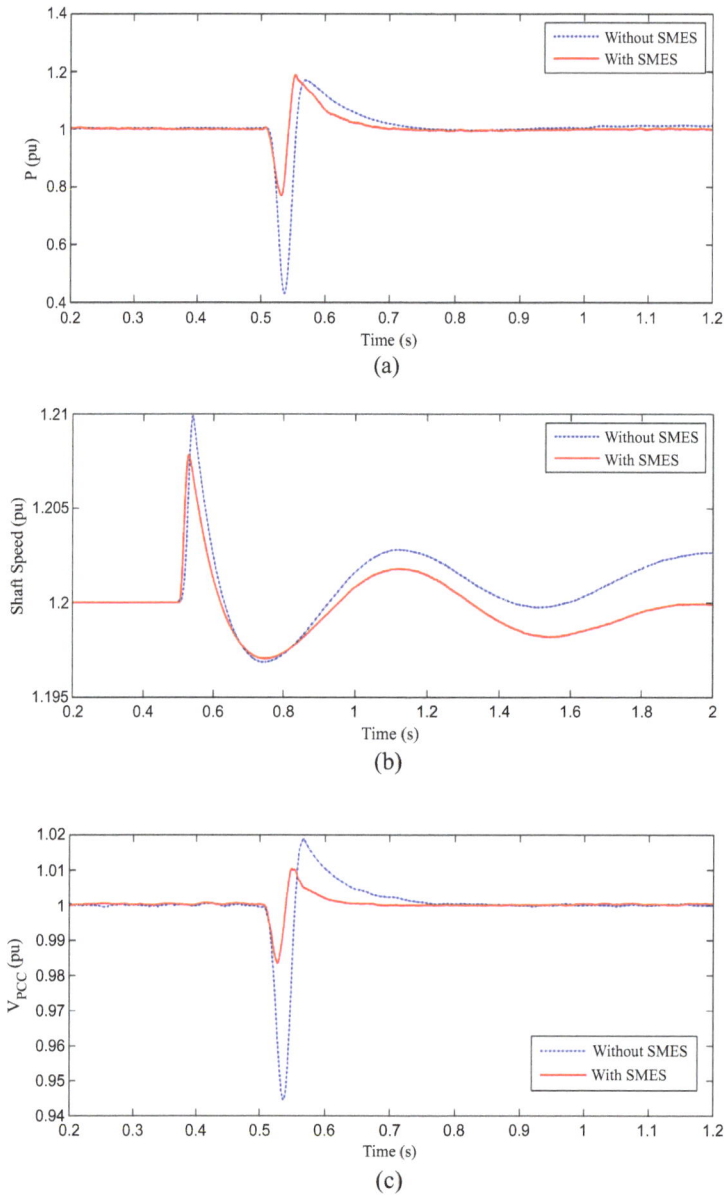

Fig. (3.28). Effect of RSC misfireon DFIG dynamic responses with and without SMES unit; (a) Power, (b) Shaft speed, and (c) Voltage at PCC.

Fire-Through on GSC and RSC

Fig. (**3.29**) shows the dynamic response of the studied system when fire-through is applied to the GSC. As shown in this figure, without SMES the dispatched power will be dropped to 0.1 pu and it takes 0.2s to settle down to the original level after fault clearance. SMES unit will slightly improve the power and correct it to 0.25 pu during the fault and it reduces the settling time.

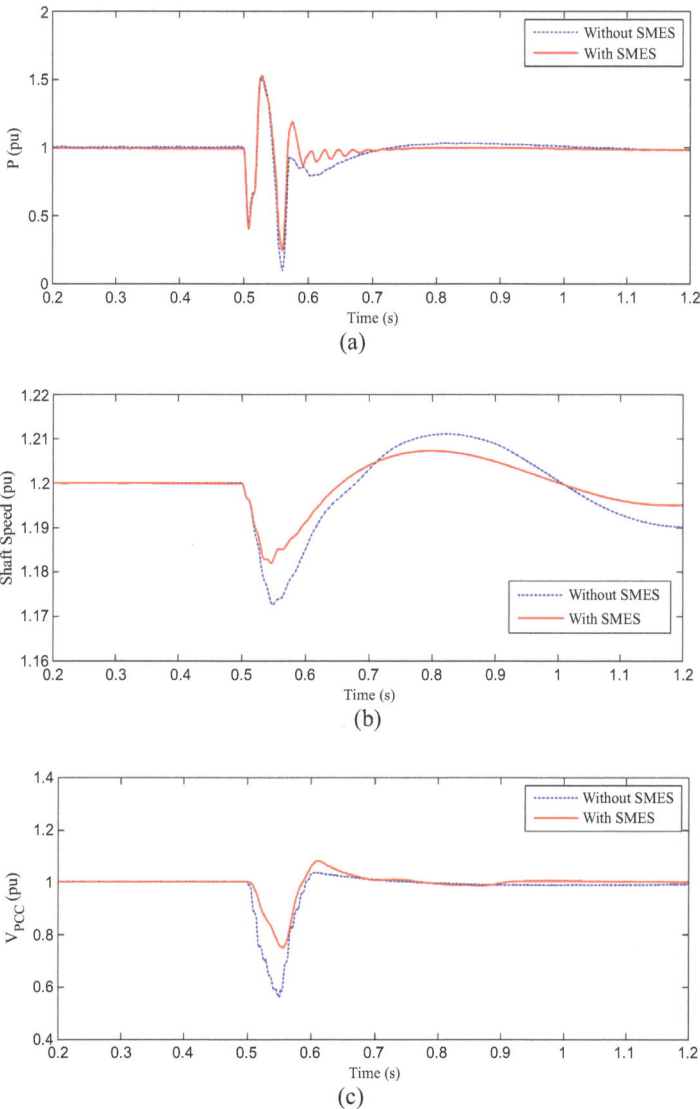

Fig. (3.29). Effect of fire-throughatGSCon DFIG's responses with and without SMES unit; (a) Power, (b) Shaft speed, (c) Electromagnetic torque and (d) Voltage at PCC.

Fig. (**3.29b**) shows that with SMES unit, shaft speed oscillation is reduced and settling time is decreased after the clearance of the fault, thus it will reach steady condition faster than the system without SMES. Furthermore, the voltage at the PCC is also improved from 0.6 pu during fault without the connection of the SMES unit to 0.8 pu when SMES is connected to the system as shown in Fig. (**3.29c**).

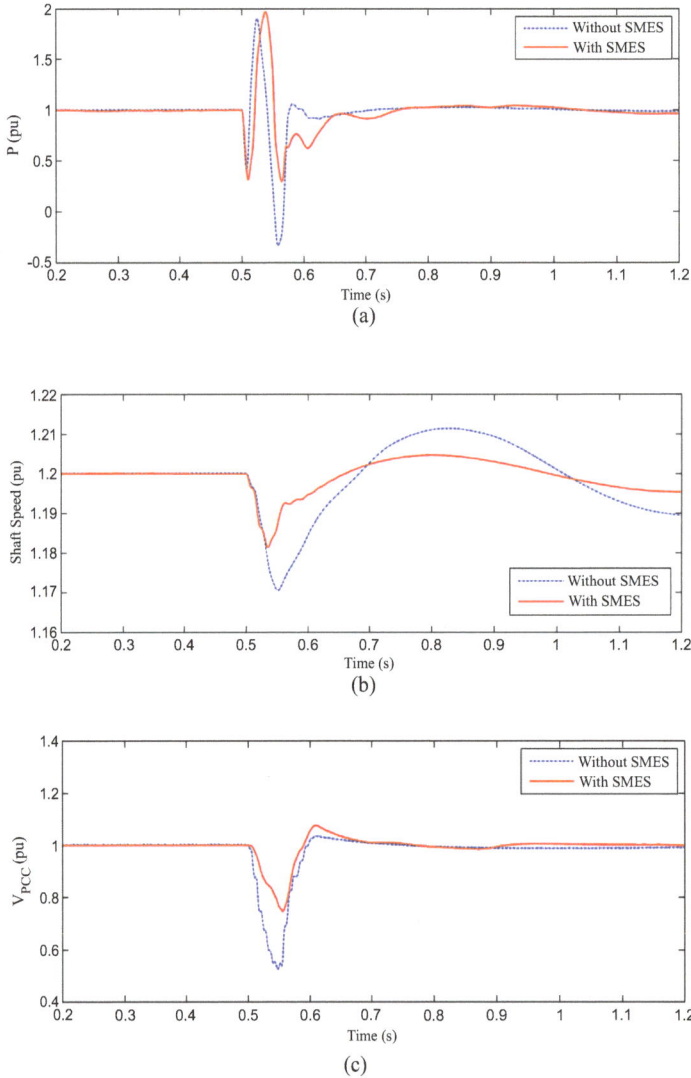

Fig. (3.30). Effect of fire-through tRSCkon DFIG's responses with and without SMES coil; (a) Power, (b) Shaft speed and (c) Voltage at PCC.

Compared to other cases, fire-thtough when occurs within RSC is the worst. Without SMES unit connected to the system, generated power oscillates and drops to negative level where the machine absorbs power from the grid and acts as a motor as shown in Fig. (**3.30a**). In this condition, protection devices such as crowbar will be activated to isolate the WTG and to protect the converter stations against excessive current. However, with the SMES unit connected to the system, the drop in generated power is modulated to 0.25 pu as shown in Fig. (**3.30a**). Both shaft speed and V_{PCC} are also significantly improved by the connection of the SMES unit as shown in Figs. (**3.30b and c**) respectively.

SUMMARY

The concept of SMES unit along with various possible control schemes has been covered in this chapter. The capability of the SMES unit in improving system stability, power quality and improving the fault ride through capability of DFIG-based WECS has been demonstrated through case studies. Although the cost of SMES unit is too high when compared with other FACTS devices, the technical benefit gained of using it in power systems is undoubtable. It is expected that the price of SMES unit will come down with the advance in low temperature superconductors that utilize liquid Nitrogen for cooling which is much cheaper than liquid Helium used for high temperature superconductors.

REFERENCES

[1]   H. Polinder, D-J. Bang, H. Li, and Z. Cheng, *Concept Report on Generator Topologies, Mechanical and Electromagnetic Optimization.* Project Up Wind, 2007.

[2]   J.M. Carrasco, L.G. Franquelo, J.T. Bialasiewicz, E. Galvan, R.C. Guisado, M.A. Prats, J.I. Leon, and N. Moreno-Alfonso, "Power-Electronic Systems for the Grid Integration of Renewable Energy Sources: A Survey", *IEEE Trans. Ind. Electron.,* vol. 53, pp. 1002-1016, 2006. [http://dx.doi.org/10.1109/TIE.2006.878356]

[3]   P.W. Carlin, A.S. Laxson, and E.B. Muljadi, *The History and State of the Art of Variable-Speed Wind Turbine Technology.* NREL: Colorado, 2001. [http://dx.doi.org/10.2172/776935]

[4]   M. Altin, O. Goksu, R. Teodorescu, P. Rodriguez, B.B. Jensen, and L. Helle, "Overview of Recent Grid Codes for Wind Power Integration", *12th International Conference on Optimization of Electrical and Electronic Equipment (OPTIM),* 2010, pp. 1152-1160 [http://dx.doi.org/10.1109/OPTIM.2010.5510521]

[5]   S. Seman, J. Niiranen, and A. Arkkio, "Ride-Through Analysis of Doubly Fed Induction Wind-Power Generator under Unsymmetrical Network Disturbance", *IEEE Trans. Power Syst.,* vol. 21, pp. 1782-1789, 2006. [http://dx.doi.org/10.1109/TPWRS.2006.882471]

[6]   J. Lopez, E. Gubia, E. Olea, J. Ruiz, and L. Marroyo, "Ride Through of Wind Turbines with Doubly Fed Induction Generator under Symmetrical Voltage Dips", *IEEE Trans. Ind. Electron.,* vol. 56, pp. 4246-4254, 2009. [http://dx.doi.org/10.1109/TIE.2009.2028447]

[7]   M. Mohseni, S.M. Islam, and M.A. Masoum, "Impacts of Symmetrical and Asymmetrical Voltage

Sags on DFIG-Based Wind Turbines Considering Phase-Angle Jump, Voltage Recovery, and Sag Parameters", *IEEE Trans. Power Electron.,* vol. 26, pp. 1587-1598, 2011.
[http://dx.doi.org/10.1109/TPEL.2010.2087771]

[8]   Y. Xiangwu, G. Venkataramanan, P.S. Flannery, W. Yang, D. Qing, and Z. Bo, "Voltage-Sag Tolerance of DFIG Wind Turbine with a Series Grid Side Passive-Impedance Network", *IEEE Trans. Energ. Convers.,* vol. 25, pp. 1048-1056, 2010.
[http://dx.doi.org/10.1109/TEC.2010.2054097]

[9]   S. Hu, X. Lin, Y. Kang, and X. Zou, "An Improved Low-Voltage Ride-Through Control Strategy of Doubly Fed Induction Generator During Grid Faults", *IEEE Trans. Power Electron.,* vol. 26, pp. 3653-3665, 2011.
[http://dx.doi.org/10.1109/TPEL.2011.2161776]

[10]   J.G. Slootweg, S.W. de Haan, H. Polinder, and W.L. Kling, "General Model for Representing Variable Speed Wind Turbines in Power System Dynamics Simulations", *IEEE Trans. Power Syst.,* vol. 18, pp. 144-151, 2003.
[http://dx.doi.org/10.1109/TPWRS.2002.807113]

[11]   F. Blaabjerg, and Z. Chen, *Power Electronics for Modern Wind Turbines.* Morgan & Claypool Publishers: Aalborg, 2006.

[12]   J. Hee-Yeol, A.R. Kim, K. Jae-Ho, P. Minwon, Y. In-Keun, K. Seok-Ho, S. Kideok, K. Hae-Jong, S. Ki-Chul, T. Asao, and J. Tamura, "A Study on the Operating Characteristics of SMES for the Dispersed Power Generation System", *IEEE Trans. Appl. Supercond.,* vol. 19, pp. 2028-2031, 2009.
[http://dx.doi.org/10.1109/TASC.2009.2018495]

[13]   S.S. Chen, L. Wang, W.J. Lee, and Z. Chen, "Power Flow Control and Damping Enhancement of A Large Wind Farm Using A Superconducting Magnetic Energy Storage Unit", *Renewable Power Generation, IET,* vol. 3, pp. 23-38, 2009.
[http://dx.doi.org/10.1049/iet-rpg:20070117]

[14]   M.H. Ali, P. Minwon, Y. In-Keun, T. Murata, and J. Tamura, "Improvement of Wind-Generator Stability by Fuzzy-Logic-Controlled SMES", *IEEE Trans. Ind. Appl.,* vol. 45, pp. 1045-1051, 2009.
[http://dx.doi.org/10.1109/TIA.2009.2018901]

[15]   F. Zhou, G. Joos, C. Abbey, L. Jiao, and B.T. Ooi, "Use of Large Capacity SMES to Improve the Power Quality and Stability of Wind Farms", *In Power Engineering Society General Meeting IEEE,* vol. 2, 2004, pp. 2025-2030
[http://dx.doi.org/10.1109/PES.2004.1373231]

[16]   M.R. Sheikh, S.M. Muyeen, R. Takahashi, T. Murata, and J. Tamura, "Minimization of Fluctuations of Output Power and Terminal Voltage of Wind Generator by Using STATCOM/SMES", In: *In PowerTech IEEE Bucharest* , 2009, pp. 1-6.
[http://dx.doi.org/10.1109/PTC.2009.5282099]

[17]   T. Asao, R. Takahashi, T. Murata, J. Tamura, M. Kubo, A. Kuwayama, and T. Matsumoto, "Smoothing Control of Wind Power Generator Output by Superconducting Magnetic Energy Storage System", *International Conference on Electrical Machines and Systems, 2007, ICEMS,* 2007, pp. 302-307

[18]   J. Hee-yeol, P. Dae-Jin, S. Hyo-Ryong, P. Minwon, and Y. In-Keun, "Power Quality Enhancement of Grid-Connected Wind Power Generation System by SMES", In: *In Power Systems Conference and Exposition, PSCE '09. IEEE/PES* , 2009, pp. 1-10.

[19]   A.M. Shiddiq-Yunus, M.A. Masoum, and A.A. Siada, "Application of SMES to Enhance the Dynamic Performance of DFIG During Voltage Sag and Swell", *IEEE Trans. Appl. Supercond.,* vol. 22, p. 5702009, 2012.
[http://dx.doi.org/10.1109/TASC.2012.2191769]

[20]   M.H. Ali, W. Bin, and R.A. Dougal, "An Overview of SMES Applications in Power and Energy Systems", *IEEE Transactions on Sustainable Energy,* vol. 1, pp. 38-47, 2010.

[http://dx.doi.org/10.1109/TSTE.2010.2044901]

[21]   I.D. Hassan, R.M. Bucci, and K.T. Swe, "400 MW SMES Power Conditioning System Development and Simulation", *IEEE Trans. Power Electron.,* vol. 8, pp. 237-249, 1993.
[http://dx.doi.org/10.1109/63.233279]

[22]   H. Li, and M.M. Gupta, *Fuzzy Logic and Inteligent System.* Kluwer Academic Publisher: Massachusetts, 1995.

[23]   A.N.S. (ANSI), IEEE Recommended Practice for Monitoring Electric Power Quality, 1995.

[24]   E.F. Fuchs, and M.A. Masoum, *Power Quality in Power Systems and Electrical Machines.* Elsevier, 2008.

[25]   V. Ahkmatov, "Analysis of Dynamic Behaviour of Power Systems with Large Amount of Wind Power", http://www.dtu.dk/upload/centre/cet/projekter/99-05/05-va-thesis.pdf, accessed: 25 February, 2011.

[26]   www.aemo.com.au, accessed: 31 December, 2011

[27]   T. Ackermann, *Wind Power in Power System.* John Wiley and Sons, Ltd.: West Sussex, 2005.
[http://dx.doi.org/10.1002/0470012684]

[28]   "Final Report System Disturbances on 4 November 2006", https://www.entsoe.eu/ fileadmin/ user_upload/ _library/ publications/ce/otherreports/Final-Report-20070130.pdf, accessed: 22 February, 2011.

[29]   T. Hamajima, H. Amata, T. Iwasaki, N. Atomura, M. Tsuda, D. Miyagi, T. Shintomi, Y. Makida, T. Takao, K. Munakata, and M. Kajiwara, "Application of SMES and Fuel Cell System Combined with Liquid Hydrogen Vehicle Station to Renewable Energy Control", In: *IEEE Transactions on Applied Superconductivity,* , 2011, pp. 1-1.

[30]   A. M. Shiddiq Yunus, A. Abu Siada, M. A. S. Masoum,, "Improving Dynamic Performance of Wind Energy Conversion System using Fuzzy-Based Hysteresis Current Controlled SMES", *Power Electronics, IET,* vol. 5, pp. 1305-1314, 2012.
[http://dx.doi.org/10.1049/iet-pel.2012.0135]

[31]   P. Boussean, P. Gautier, E. Garzulino, I. Juston, and R. Belhomme, *Grid Impact of Different Technologies of Wind Turbine Generator Systems.* EDF Electricite de France, ClamartCedex, 2003.

[32]   J. Arrillaga, *High Voltage Direct Current Transmission.* Peter Peregrinus, Ltd.: London, 1983.

[33]   K.R. Padiyar, *HVDC Power Transmission Systems.* John Wiley & Sons: New Delhi, 1990.

[34]   L. Bin, and S. Sharma, "A Survey Of IGBT Fault Diagnostic Methods for Three-Phase Power Inverters", *International Conference on Condition Monitoring and Diagnosis,* 2008, pp. 756-763
[http://dx.doi.org/10.1109/CMD.2008.4580396]

[35]   W.L. Fuchs, and E.F. Fuchs, "Frequency Variations of Power System Due to Switching of Renewable Energy Sources", *International Conference on Renewable Energies and Power Quality (ICREPQ'12)* Santiago de Compostela, Spain, p.6. 2012.
[http://dx.doi.org/10.24084/repqj10.246]

[36]   R.K. Behera, and G. Wenzhong, "Low Voltage Ride-Through and Performance Improvement of a Grid Connected DFIG System", *International Conference on Power Systems,* 2009, pp. 1-6
[http://dx.doi.org/10.1109/ICPWS.2009.5442737]

[37]   S. Hu, and H. Xu, "Experimental Research on LVRT Capability of DFIG WECS during Grid Voltage Sags", *Power and Energy Engineering Conference,* 2010, pp. 1-4
[http://dx.doi.org/10.1109/APPEEC.2010.5448151]

[38]   K. Lima, A. Luna, E.H. Watanabe, and P. Rodriguez, "Control strategy for the Rotor Side Converter of A DFIG-WT Under Balanced Voltage Sag", *Power Electronics Conference,* 2009, pp. 842-847
[http://dx.doi.org/10.1109/COBEP.2009.5347615]

[39]   W. Yulong, L. Jianlin, H. Shuju, and X. Honghua, "Analysis on DFIG Wind Power System Low-Voltage Ride-Through", *International Joint Conference on Artificial Intelligence,* 2009, pp. 676-679

[40]   A.M. Shiddiq-Yunus, A. Abu-Siada, and M.A. Masoum, "Improvement of LVRT Capability of Variable Speed Wind Turbine Generators Using SMES Unit", Innovative Smart Grid Technologies (ISGT) Asia, IEEE PES, Perth, Western Australia. 2011.

[41]   L. Trilla, O. Gomis-Bellmunt, A. Junyent-Ferre, M. Mata, J. Sanchez, and A. Sudria-Andreu, "Modeling and Validation of DFIG 3 MW Wind Turbine Using Field Test Data of Balanced And Unbalanced Voltage Sags", In: *IEEE Transactions on Sustainable Energy*, 2011, , pp. 1-1.

[42]   A. Abu-Siada, and S. Islam, "Application of SMES Unit in Improving the Performance of an ac/dc Power System", *IEEE Transactions on Sustainable Energy,* vol. 2, pp. 109-121, 2011.
[http://dx.doi.org/10.1109/TSTE.2010.2089995]

[43]   S.O. Faried, and A.M. El-Serafi, "Effect of HVDC converter Station Faults on Turbine-Generator Shaft Torsional Torques", *IEEE Trans. Power Syst.,* vol. 12, pp. 875-881, 1997.
[http://dx.doi.org/10.1109/59.589740]

[44]   H.A. Darwish, A.M. Taalab, and M.A. Rahman, "Performance of HVDC Converter Protection during Internal Faults", in Power Engineering Society General Meeting, 2006, pp. 7. 2006.
[http://dx.doi.org/10.1109/PES.2006.1709425]

[45]   H.B. Sethom, and M.A. Ghedamsi, "Intermittent Misfiring Default Detection and Localisation on a PWM Inverter Using Wavelet Decomposition", *Journal of Electrical System,* vol. 4, pp. 222-234, 2008.

[46]   A.M. Shiddiq-Yunus, A.A. Siada, and M.A. Masoum, "Impact of SMES on DFIG Power Dispatch during Intermittent Misfire and Fire-through Faults", *IEEE Trans. Appl. Supercond.,* vol. 23, p. 5701712, 2013.
[http://dx.doi.org/10.1109/TASC.2013.2256352]

Recent Advances in Renewable Energy, 2017, *Vol. 2*, 87-141

CHAPTER 4

Distribution Static Compensators and their Applications in Microgrids

Abstract: Microgrids are clusters of distributed energy resources, energy storage systems and loads which are capable of operating in grid-connected as well as in off-grid modes. In the off-grid mode, the energy resources supply the demand while maintaining the voltage and frequency within acceptable limits whereas in the grid-connected mode, the energy resources supply the maximum or nominal power and the network voltage and frequency is maintained by the grid. This chapter first summarizes the structure and control principles of microgrids. It then briefly introduces the structures and control perspectives of distribution static compensators (DSTATCOMs). Finally, some applications of DSTATCOMs are discussed in microgrids. The introduced applications are power quality improvement due to the presence of nonlinear and unbalanced loads, voltage regulation and balancing, and interphase power circulation in the case of the presence of single-phase energy resources with unequal distribution amongst phases. Each application is illustrated by examples, realized in PSCAD/EMTDC.

Keywords: Distribution static compensator (DSTATCOM), Interphase power circulation, Load compensation, Microgrid, Voltage regulation.

INTRODUCTION

The growing electricity demand, the necessity of electricity generation-related costs and emissions, on top of the necessity for improving the reliability of electrical systems and reducing its losses, are driving the electricity utilities around the world towards the utilization of distributed energy resources (DER) within distribution networks and their installation at close distances to the consumption centers [1].

With the introduction of DERs installation in distribution networks and at proximity of loads, the concept of microgrids has been introduced. Microgrids are systems with clusters of DERs, energy storage systems, and loads [2]. Fig. (**4.1**) illustrates schematically a microgrid system. To deliver power with high quality and reliability, the microgrid should appear as a controllable system that responds punctually to the demand and/or generation changes. Microgrids can operate in a

Ahmed Abu-Siada, Mohammad A.S. Masoum, Yasser Alharbi, Farhad Shahnia & A.M. Shiddiq Yunus

grid-connected mode as well as in off-grid (also known as islanded or autonomous) mode. The off-grid mode of operation can be during network maintenance (a planned islanding of the microgrid) or during faults in the network (an unplanned islanding). However, some microgrids, especially those in remote areas, usually referred to as standalone microgrids, operate in off-grid mode permanently. Smaller microgrids, also referred to as nanogrids, are also thought to be the principle of future power supply systems to residential customers [3]. In grid-connected mode, the grid dictates the voltage and frequency of the microgrid and the DERs within the microgrid operate at their nominal capacities (for dispatchable ones) or based on their maximum power point tracking (for renewable energy-based non-dispatchable ones). Thus, the DERs are usually controlled under a constant PQ control strategy in this mode of operation. However, in the off-grid mode, one or a group of the DERs control the voltage and frequency, in addition to sharing the load demand. Different control techniques can be used for the operation of DERs in this mode of operation. Some of them can be simple and decentralized (without a need for any communication system) while some can be more complex and comprehensive to yield optimal results (if proper data communication systems are in place) [4]. Voltage and frequency droop control is one of the simplest while effective decentralized techniques for the control of DERs [5]. Also, droop control can result in power sharing among multiple DERs in the microgrid similar to the power sharing among synchronous generators in traditional power systems. Voltage and angle droop can also be used instead of the voltage and frequency droop [6, 7]. A few modified or intelligent droop controls are also proposed to either improve the system performance or stability [8]. Once the references for the DERs are assigned by the droop control, the DERs are controlled through their primary controls to track the desired references in their outputs.

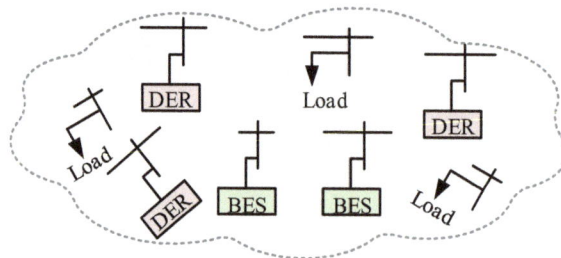

Fig. (4.1). Schematic of a microgrid consisting of DERs, battery energy storage (BES) systems, and loads.

On the other hand, one of the main concerns in any electrical system including microgrids is the reliability of power supply which is defined by different indices to express the system's or its customer's average interruption frequency or duration [9]. However, reliability is not the only concern these days as power quality is equally important. Power quality is expressed as having a supply voltage with a pure sine curve, with the nominal magnitude and frequency. However, small deviations such as ± 5% in these quantities are often acceptable [10]. The reasons of power quality problems are generally complex and vary from load characteristics, DER and converter characteristics to network component features. Power quality problems have different adverse effects. As an example, an impulsive transient can cause damage to some equipment with insulations such as transformers and cables. Transient overvoltages can damage power line insulators. Undervoltages can cause tripping of the electrical motors or electronic equipment while overvoltages can cause damage to a large group of electrical devices. Voltage unbalance may result in a temperature rise and tripping in electrical motors. Harmonics also can lead to the malfunction of ripple control or traffic control systems, as well as increased losses in power transformers. Unwanted harmonic currents flowing in the distribution lines can increase the network losses. DC offsets can result in the saturation of power transformers.

Custom power devices can improve the power quality of an electrical system and can minimize the presence and spread of under/over-voltages, harmonics, unbalance, *etc.* [11, 12]. They are power electronics-based converters that are used in distribution networks and can eliminate the undesired power quality problems from affecting the customers. Some custom power devices are used for network reconfiguration (supplying a load from multiple feeders that can be switched depending on the quality or the availability of the power in each feeder) while some are used for load compensation such as distribution static compensators (DSTATCOMs) [13]. A DSTATCOM is a shunt-connected custom power device that has been traditionally used for power factor correction, harmonic filtering, and current balancing (referred to as load compensation) or balancing the voltage at its connection point (referred to as voltage regulation). Microgrids are an interesting application point for DSTATCOMs due to several reasons:

1- Power quality is an important issue in every electrical system including microgrids; thereby, DSTATCOMs can be used to improve the power quality in microgrids. Especially, they can compensate the harmonics induced by the converters of renewable energy-based DERs or improve the nonlinear and power factor problems of the loads [11]. This falls under the traditional performance of DSTATCOMs.

2- Most of the renewable energy-based DERs are connected to the microgrid feeder through a power electronics-based converter, mostly a three-phase one. There is a large similarity between the structure of these converters and the one of a DSTATCOM while the difference is in their ratings and control techniques. The power converter of a renewable energy-based DER can be modified to not only inject the generated power by the renewable source to the feeder but also improve the power quality of the microgrid [14]. This is a combination of DSTATCOM and renewable energy-based DER with the functionalities of both of them.

3- The research on MGs has been mostly focused on three-phase ac MGs, dc MGs or their hybrid combinations. So far, not enough research is carried out on the operation of single-phase DERs within three-phase MGs, specifically when the MG is operating in autonomous mode [15]. DSTATCOMs can be very helpful in this aspect as they can provide a path for the circulation of excess power from one phase to the other phases, when the single-phase DERs are distributed unequally and randomly amongst phases.

With the above introduction, this chapter first summarizes the structure and control principles of microgrids. Then, briefly it introduces the structures and control perspectives of DSTATCOMs. Finally, the above-mentioned applications of DSTATCOMs are discussed in microgrids while each application is illustrated by an example that is realized in PSCAD/EMTDC.

MICROGRID STRUCTURE AND CONTROL

For appropriate operation of a microgrid within a distribution network, a three-level hierarchical control system is required [16]; which is discussed below and illustrated schematically in Fig. (**4.2**).

- The primary controller comprises of inner-loop and outer-loop controls. Inner-loop control is responsible for appropriate operation of the DER such that a proper tracking of the desired reference is achieved at the output of the DER. In converter-based DERs, this is achieved by appropriate switching in the converter. This control is based on the references determined by the outer-loop control and the local current, voltage and power measurements at the DER converter output. Outer-loop control facilitates proper output power control of DERs in the microgrid. This control generates the appropriate references for the inner-loop control which is different for grid-connected and autonomous modes [8, 16].
- The secondary control is the central controller of the microgrid. In off-grid mode of operation, the secondary controller determines the desired (reference) voltage magnitude and angle or frequency for each DER's primary controller by monitoring the network voltage and frequency. This controller runs in a slower time frame compared to that of the primary control and sends those references only if either of voltage or frequency deviate beyond the acceptable limits [17].

- The tertiary control is the network highest control. This controller communicates with the secondary controller of each microgrid on top of the protection devices and circuit breakers of the network. It can use the load forecasting, weather forecasting, electricity market, and demand response information to yield the optimal operation of the microgrids and the network [16]. If multiple microgrids exist in the network and proper links are available among them, this controller can also interconnect them appropriately to prevent voltage/frequency collapse in a microgrid during excessive loading or prevent renewable curtailment for a microgrid experiencing excessive generation from its renewable sources [18 - 20].

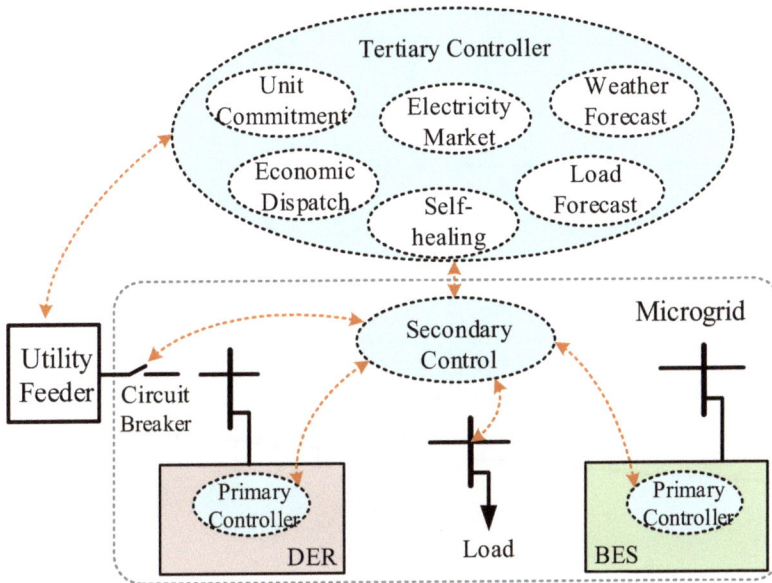

Fig. (4.2). Schematic of a microgrid with its hierarchical control.

Control of DER Systems

It is expected that in near future, the microgrids will be dominated by renewable energy-based DERs which will be connected to the microgrid through voltage source converters (VSCs) [21]. Thus, tomorrow's microgrids are thought to be sustainable, converter-dominated microgrids. Thereby, in the rest of this section, the control of converter-interfaced DERs are discussed only.

Let us consider the VSC and filter structure of Fig. (**4.3**). This circuit can be represented by its Thevenin equivalent parameters (*i.e.* V_{Th} and Z_{Th}). The instantaneous active power (p) and reactive power (q) injected from a DER to its point of common coupling (PCC) can be expressed as [16].

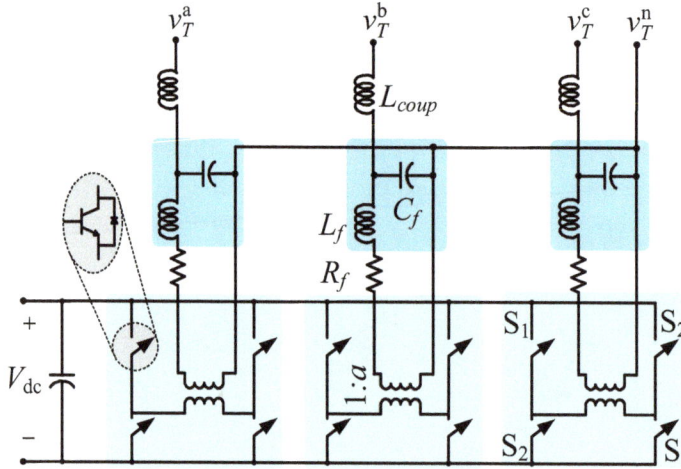

Fig. (4.3). Considered three-phase VSC and filter system for the DERs and BESs.

$$p = \frac{\left(|V_T||V_{Th}|\cos\phi - |V_{Th}|^2\right)\cos\theta + |V_T||V_{Th}|\sin\phi\sin\theta}{|Z_{Th} + Z_{coup}|}$$

$$q = \frac{\left(|V_T||V_{Th}|\cos\phi - |V_{Th}|^2\right)\sin\theta - |V_T||V_{Th}|\sin\phi\cos\theta}{|Z_{Th} + Z_{coup}|}$$

$$\text{(4.1)}$$

where $\phi = \delta_{Th} - \delta_T$ and $\theta = \angle(Z_{Th} + Z_{coup})$ in which V_T is the PCC voltage, Z_{coup} is the coupling impedance and $V = |V| \angle\delta$ represents the phasor notation of $v(t)$. The coupling impedance is dominantly inductive (*i.e.* $Z_{coup} \approx j\omega L_{coup}$). Z_{Th} is also dominantly inductive at 50 Hz (*i.e.* $Z_{Th} \approx j\omega L_{conv}$) and has a minuscule magnitude. Thus, $V_{Th} \approx V_{cf}$ where V_{cf} is the voltage across capacitor C_f. Based on these assumptions, (1) can be simplified as

$$p = \frac{|V_T||V_{cf}|\sin(\delta_{cf} - \delta_T)}{\omega L_{conv} + \omega L_{coup}}$$

$$q = \frac{|V_T||V_{cf}|\cos(\delta_{cf} - \delta_T) - |V_T|^2}{\omega L_{conv} + \omega L_{coup}}$$

$$\text{(4.2)}$$

The average active power (P) and reactive power (Q) that are injected by each DER are calculated from p and q by the help of a low pass filter.

Now, consider the microgrid of Fig. (**4.4**) with 2 DERs and one load. The DERs are connected to the load through feeder impedances $Z_{line,1}$ and $Z_{line,2}$. The feeder is assumed to be highly inductive ($Z_{line} \approx j\omega L_{line}$). Assuming the voltage at load PCC as V_{load}, the active and reactive powers injected by each DER are

$$p = \frac{|V_{load}||V_{cf}|\sin(\delta_{cf} - \delta_{load})}{\omega L_{conv} + \omega L_{coup} + \omega L_{line}}$$

$$q = \frac{|V_{load}||V_{cf}|\cos(\delta_{cf} - \delta_{load}) - |V_{load}|^2}{\omega L_{conv} + \omega L_{coup} + \omega L_{line}}$$

(4.3)

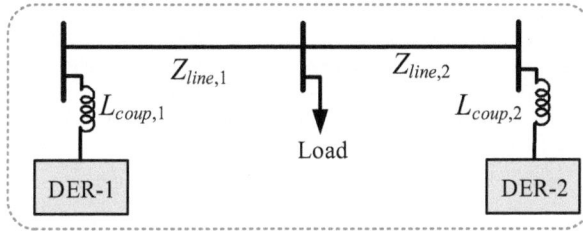

Fig. (**4.4**). A microgrid composed of 2 DERs and one load.

As the feeder is highly inductive, the active and reactive powers are decoupled, and a dc load flow analysis can be applied. Also, the angle difference between V_{cf} and V_{load} is small. The average active power output of each DER, according to the dc load flow, becomes

$$P = \frac{|V_{load}||V_{cf}|(\delta_{cf} - \delta_{load})}{\omega L_{conv} + \omega L_{coup} + \omega L_{line}}$$

(4.4)

To supply the load with an average active power of P, the angle of the voltage across the filter capacitor of the DER has to be

$$\delta_{cf} = P(\Gamma_{conv} + \Gamma_{coup} + \Gamma_{line}) + \delta_{load}$$

(4.5)

where

$$\Gamma = \frac{\omega L}{\left|V_{cf}\right|\left|V_{load}\right|}.$$

Power sharing among multiple DERs in a microgrid can be achieved by changing the voltage magnitude and angle of each DER, using the droop control, as

$$\delta_{cf} = \delta_{rated} - \frac{m}{Z_{line}}\left[X_{line}\left(P - P_{rated}\right) - R_{line}\left(Q - Q_{rated}\right)\right]$$

$$\left|V_{cf}\right| = V_{rated} - \frac{n}{Z_{line}}\left[R_{line}\left(P - P_{rated}\right) + X_{line}\left(Q - Q_{rated}\right)\right] \tag{4.6}$$

where V_{rated} and δ_{rated} are the rated voltage magnitude and angle of the DER respectively when it injects its rated active and reactive powers (P_{rated}, Q_{rated}). Active power-angle and reactive power-voltage droop coefficients are represented by m in rad/Ws and n in V/VAr, respectively. As the lines are assumed inductive, (6) can be simplified as

$$\delta_{cf} = \delta_{rated} - m\left(P - P_{rated}\right)$$

$$\left|V_{cf}\right| = V_{rated} - n\left(Q - Q_{rated}\right) \tag{4.7}$$

This is monitored and controlled in the microgrid through the outer-loop control.

When using a frequency droop instead of the angle droop, the outer-loop control can be expressed for each DER as

$$f = f_{rated} - \frac{m'}{Z_{line}}\left[X_{line}\left(P - P_{rated}\right) - R_{line}\left(Q - Q_{rated}\right)\right]$$

$$\left|V_{cf}\right| = V_{rated} - \frac{n}{Z_{line}}\left[R_{line}\left(P - P_{rated}\right) + X_{line}\left(Q - Q_{rated}\right)\right] \tag{4.8}$$

$$f = f_{rated} - m'\left(P - P_{rated}\right)$$

$$\left|V_{cf}\right| = V_{rated} - n\left(Q - Q_{rated}\right) \tag{4.9}$$

instead of (6)-(7), where f_{rated} is the rated frequency of the microgrid and m' is the droop coefficient in Hz/W. Fig. (**4.5**) illustrates the droop control schematically.

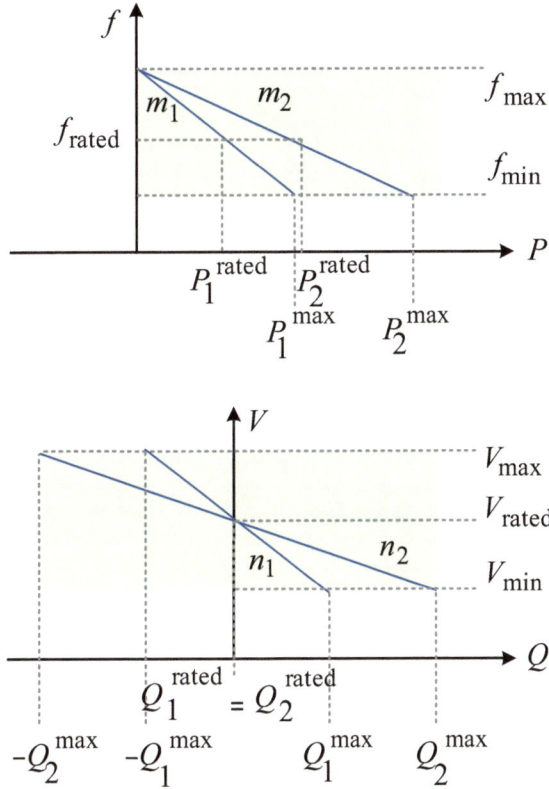

Fig. (4.5). Droop curves for two DERs in an islanded microgrid.

Assuming that a DER's frequency will reduce by $\Delta\omega$ when it increases its injected active power from zero to its rated capacity, the active power-angle droop coefficient for this DER will be

$$m = \frac{\Delta\omega}{P_{rated}}$$

(**4.10**)

and assuming $\Delta\omega$ constant for all DERs, even those with different rated capacities, for every two DER, one gets

$$\frac{m_i}{m_j} = \frac{P_{rated,j}}{P_{rated,i}}$$

(**4.11**)

Likewise, assuming that a DER's voltage will reduce by ΔV when it increases its injected reactive power from zero to its nominal capacity, the reactive power-voltage droop coefficient for this DER will be

$$n = \frac{\Delta V}{Q_{rated}} \tag{4.12}$$

and assuming ΔV constant for all DERs, even those with different rated capacities, for every two DER, one gets

$$\frac{n_i}{n_j} = \frac{Q_{rated,j}}{Q_{rated,i}} \tag{4.13}$$

Assuming DER-i and DER-j have the same δ_{rated} in the microgrid of (Fig. **4.4**) at the steady-state condition, from (7) and (11), one gets

$$\delta_{cf,i} - \delta_{cf,j} = \left(\delta_{rated,i} - \delta_{rated,j}\right) - m_i\left(P_i - P_{rated,i}\right) + m_j\left(P_j - P_{rated,j}\right) = m_j P_j - m_i P_i \tag{4.14}$$

Replacing $\delta_{cf,i}$ and $\delta_{cf,j}$ from (5) in (14) yields

$$P_i\left(\Gamma_{conv,i} + \Gamma_{coup,i} + \Gamma_{line,i}\right) - P_j\left(\Gamma_{conv,j} + \Gamma_{coup,j} + \Gamma_{line,j}\right) = m_j P_j - m_i P_i \tag{4.15}$$

Thereby, the ratio of the injected active power by the DERs will be

$$\frac{P_j}{P_i} = \frac{\Gamma_{conv,i} + \Gamma_{coup,i} + \Gamma_{line,i} + m_i}{\Gamma_{conv,j} + \Gamma_{coup,j} + \Gamma_{line,j} + m_j} \tag{4.16}$$

It is seen from (16) that the injected active power of a DER is inversely proportional to the sum of Γ in its output. Γ_{conv}, Γ_{coup} and Γ_{line} depend on the inductances between the DER and the load. Since Γ has a parameter of voltage square in its denominator, it is expected that

$$\Gamma_{conv} \ll \Gamma_{line} \ll \Gamma_{coup} \ll m \tag{4.17}$$

Hence, (16) can be rewritten as

$$\frac{P_j}{P_i} \approx \frac{m_i}{m_j} \tag{4.18}$$

In a similar way, the ratio of the reactive power injected by the DERs are

$$\frac{Q_j}{Q_i} \approx \frac{n_i}{n_j} \tag{4.19}$$

Based on the above assumption, from (11), (18) and (13), (19), it is expected that the injected active and reactive power ratio among two DERs in the microgrid will be same as the proportion of their nominal active and reactive power capacities.

On the other hand, it is desired that the voltage angle difference across the coupling inductance (*i.e.* $\delta_{cf} - \delta_{PCC}$) of all DERs to be small so that it is in the linear section of sinusoidal P-δ characteristic of (2). Similarly, it is desired that the voltage drop across the coupling inductances (*i.e.* $|V_{cf}| - |V_{PCC}|$) to be small. To realize these assumptions, the coupling inductances should be designed inversely proportional to the nominal power ratio of DERs as

$$\frac{L_{coup,i}}{L_{coup,j}} = \frac{P_{rated,j}}{P_{rated,i}} = \frac{Q_{rated,j}}{Q_{rated,i}} \tag{4.20}$$

Finally, it can be concluded that for an accurate power sharing among DERs in microgrid, it is required to have [16].

$$\frac{P_j}{P_i} \approx \frac{m_i}{m_j} = \frac{L_{coup,i}}{L_{coup,j}} = \frac{P_{rated,j}}{P_{rated,i}}$$
$$\frac{Q_j}{Q_i} \approx \frac{n_i}{n_j} = \frac{L_{coup,i}}{L_{coup,j}} = \frac{Q_{rated,j}}{Q_{rated,i}} \tag{4.21}$$

It is to be noted that (21) is valid as far as dynamic ratio change is not required throughout the operation of the microgrid. If the ratio of the injected power of the DERs may be required to change dynamically, the modified techniques of [17, 22 - 24] can be utilized.

As an example, consider the microgrid system of (Fig. **4.4**) in which it is desired to maintain the injected power by DER-1 twice of that injected by DER-2. Thereby, it is assumed that $L_{coup,1} = 1.36$ mH, $m_1 = 1.5708$ rad/kWs, $n_1 = 9$ V/kVAr while $L_{coup,2} = 2L_{coup,1}$, $m_2 = 2m_1$, and $n_2 = 2n_1$. The microgrid is assumed at the steady-state condition, with a load demand of 0.41 pu. At $t = 0.5$ s, the load is increased to 1 pu while it is reduced to 0.53 pu at $t = 1$ s. At $t = 1.5$ s, the load is further reduced to 0.17 pu. As seen from (Fig. **4.6a**), the injected active power ratio among DER-1 and 2 is maintained as 2:1 for all load changes. The load demand is shown in Fig. (**4.6b**). The microgrid voltage and frequency are within the acceptable limits during all load changes, as shown in Fig. (**4.6c-d**) [17].

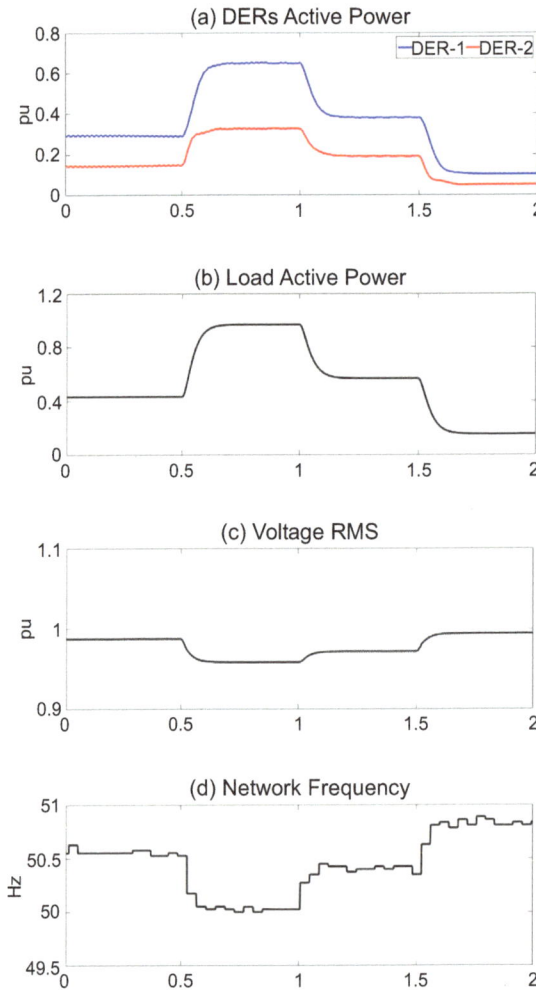

Fig. (4.6). Sample dynamic results for the considered microgrid system.

Control of BES Systems

Applying the above-discussed droop control for the BES systems of a microgrid will have some limitations. This is due to the fact that in this way, the injected power of BES systems will not be proportional to their stored energy levels. Therefore, the period that a BES system can supply a load is not controlled; thus, the BES systems with lower state of charge (SoC) will quickly run out of energy. To improve this, a modified control algorithm is preferred for the BES systems, which considers their SoC instead of their rated capacity in (21). The stored energy in the BES systems can be defined by

$$E_{\text{BES}} = E_{\text{BES}}^{\text{initial}} - \int P_{\text{BES}}\, dt \qquad (4.22)$$

where $E_{\text{BES}}^{\text{initial}}$ is the initial stored energy in the BES system and P_{BES} is its injected average active power. Thus, the SoC of the BES can be calculated from

$$SoC_{\text{BES}} = \frac{E_{\text{BES}}}{E_{\text{BES}}^{\max}} \qquad (4.23)$$

where SoC_{BES} and $E_{\text{BES}}^{\text{initial}}$ represent respectively the instantaneous SoC of the BES and its energy capacity. The droop control of the BES systems are dynamically adjusted and modified considering their SoC. As the SoC of the BES changes dynamically, a discrete value of SoC ($SoC_{\text{BES}}^{\text{d}}$), defined as

$$SoC_{\text{BES}}^{\text{d}} = \begin{cases} 1 & SoC = 100\% \\ 0.9 & 90\% \le SoC < 100\% \\ 0.8 & 80\% \le SoC < 90\% \\ \vdots & \\ 0.2 & 20\% \le SoC < 30\% \end{cases} \qquad (4.24)$$

can be applied in the droop control. The BES system will stop to discharge if SoC_{BES} drops to its minimum (*e.g.* 20% of $E_{\text{BES}}^{\text{initial}}$).

For each BES system, the P-δ droop coefficient is defined from [25]

$$\frac{P_{\text{BES}}}{P_0} = \frac{SoC^{\text{d}}_{\text{BES}}}{SoC_0} = \frac{m_0}{m_{\text{BES}}} \tag{4.25}$$

where m_0 is the droop coefficient designed for P_0 and $SoC_0 = 1$. Using (25), the droop coefficient of a BES is defined as

$$m_{\text{BES}} = \frac{m_0}{SoC^{\text{d}}_{\text{BES}}} \tag{4.26}$$

which varies over time. As a result of these actions, the BES system with a smaller SoC injects a smaller active power which avoids its fast discharge. This is illustrated schematically in Fig. (**4.7**).

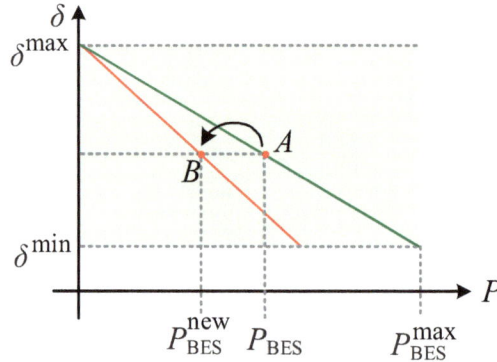

Fig. (4.7). Dynamic SoC-based droop coefficient adjustment in BES.

Thereby, in a microgrid consisting of two BES systems, their injected power ratio is as

$$\frac{P_{\text{BES1}}}{P_{\text{BES2}}} = \frac{SoC_{\text{BES1}}}{SoC_{\text{BES2}}} = \frac{m_{\text{BES2}}}{m_{\text{BES1}}} \tag{4.27}$$

The limitation of this approach is that the injected power of a BES system will change whenever the injected power of another BES system reduces. To amend this, the droop coefficients of the DERs can be updated as

$$m_{\text{DER}} = \frac{m_0}{\displaystyle\prod_{j=1}^{N} SoC^{\text{d}}_{\text{BES-}j}} \tag{4.28}$$

where N represents the number of the BES systems within the microgrid. This is schematically shown in Fig. (**4.8**). As seen from this figure, following the decrease in the SoC of one BES (which is accompanied by the relocation of the operating point of that BES from point-A to B), the operating point of the DER relocates from point-C to D. Consequently, the DER increases its injected power. Under such a case, the output power of each BES becomes only correlated to its own SoC, and it is independent of the SoC of the rest of BES systems [25]. The main disadvantage of this technique is the necessity of a communication system that transmits the SoC of the BES systems to the DERs.

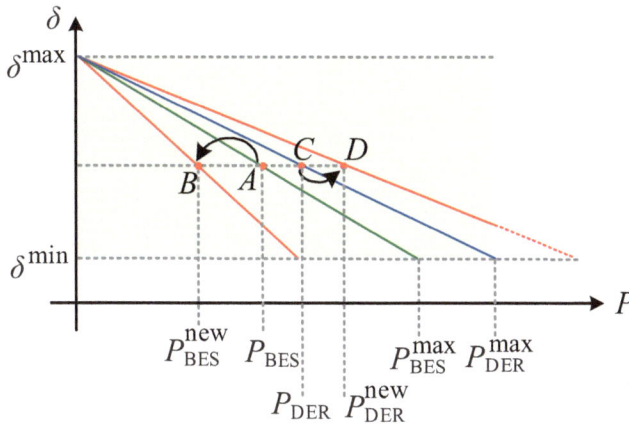

Fig. (4.8). Simultaneous droop coefficient adjustment in DERs and BESs.

As seen from (27), the injected active power of a BES reduces when (SoC_{BES}^{d}) becomes smaller. Thus, a smaller share of the capacity of the converter of the BES (S_{BES}^{cap}) is used for active power flow. The unused share can then be used for reactive power exchange. Thereby, in addition to active power exchange between the BES system and the microgrid, a reactive power exchange can also occur. The available capacity in the converter of a BES (Q_{BES}^{av}) can be determined as

$$Q_{BES}^{av} = \sqrt{\left(S_{BES}^{cap}\right)^2 - \left(P_{BES}\right)^2}\qquad(4.29)$$

and then used to defined the voltage-reactive power droop coefficient of BES (n_{BES}) as

$$n_{BES} = \frac{\Delta V}{Q_{BES}^{av}}\qquad(4.30)$$

It is noteworthy that the reactive power exchange between the converter of the BES system and the microgrid does not affect the stored energy nor the SoC of the BES system.

As an example, consider the microgrid system of Fig. (**4.9**) with two DERs and two BES systems. The maximum power capacity of DER-1 is assumed twice of that of DER-2 while the maximum energy capacities of BES-1 and BES-2 are thought to be equal. Fig. (**4.10a**) shows the load demand. Applying the modified droop technique for the BESes, it can be seen from Fig. (**4.10b**) that when the SoC of one BES decreases, it reduces its injected active power while the injected active power of the DERs increases. The SoC variations of the BES systems are shown in Fig. (**4.10c**) [25].

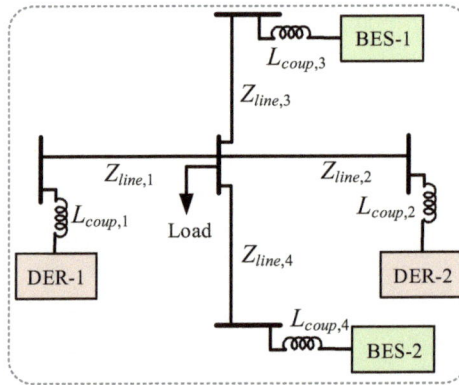

Fig. (4.9). Considered microgrid system with two DERs and two BES systems.

DSTATCOM STRUCTURE AND CONTROL

As discussed in the Introduction Section, DSTATCOM is a shunt-connected VSC-based device. There are a large range of VSCs that can be used for this purpose. The usual practice is to use full-bridge converters with three or four legs. However, depending on the power levels, multilevel and multi modular ones can also be used [13]. Each converter also requires passive filters in the output to eliminate high frequency switching harmonics from its output current and voltage waveforms.

In addition, different techniques can be used for VSC modulation such as the pulse width modulation (PWM), hysteresis, and space vector, as well as the output current and voltage control [13]. This section briefly describes the most well-known structures of VSCs, passive filters, and modulation and control techniques that can be used in DSTATCOMs.

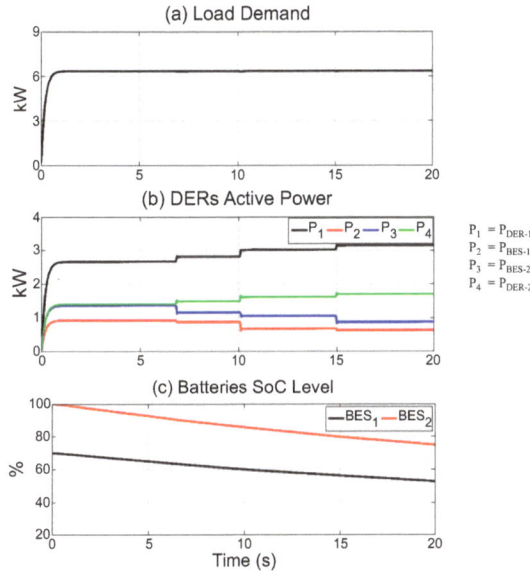

Fig. (4.10). Sample results for the considered microgrid system.

VSC Structures

Fig. (**4.11**) illustrates the schematic diagram of a three-phase, three-wire, three-leg, full-bridge VSC. This is the most common form of a three-phase converter and is used in different applications. The dc side of the converter (the dc link) has a capacitor of C_{dc} to regulate the voltage at its input and has a voltage of V_{dc}. The converter contains six switches, each consisting of a power semiconductor device such as insulated gate bipolar transistor (IGBT), metal-oxide-semiconductor field-effect transistor (MOSFET), etc., snubber circuit, and an anti-parallel diode to facilitate the continuity of current when the switch turns off. The switches in each leg operate complementarily to avoid the short-circuiting of the dc link by turning on both switches of a leg simultaneously. The disadvantage of this VSC is that the algebraic sum of the three output currents must be zero; thus, it is not suitable for unbalanced loads. In such a condition, only two phases can be controlled independently [26].

Fig. (**4.12**) shows a neutral-clamped three-phase, full-bridge VSC. Its dc link is composed of two identical series-connected capacitors, with a center point which is directly grounded (in the case of three-phase three-wire system) or connected to the existing neutral wire (in the case of three-phase, four-wire systems). Each capacitor holds a voltage of $V_{dc}/2$; thus, the dc link voltage is equal to V_{dc}. In the case of unbalanced loads, the center point of the dc link provides a path for the zero sequence of the current to flow; and thereby, all legs of the VSC can be controlled individually [26].

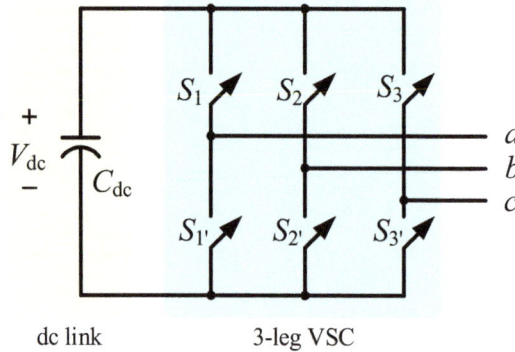

Fig. (4.11). A three-phase, three-leg, three-wire full-bridge VSC.

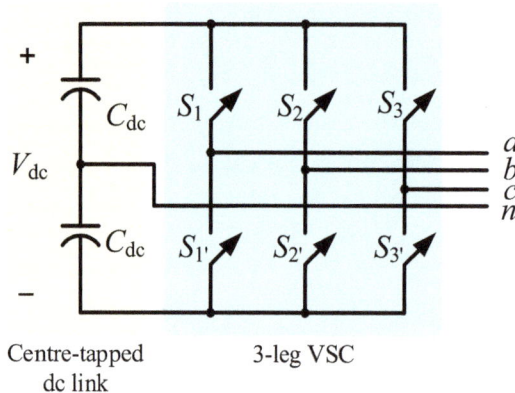

Fig. (4.12). A neutral-clamped, three-phase, full-bridge VSC.

Fig. (**4.13**) shows a three-phase, four-leg VSC in which the center point of leg-4 will connect to the neutral point through a resistor and an inductor and the current flowing through this path cancels the zero-sequence component of the current of the unbalanced loads. Thereby, similar to the previous VSC, all legs can be controlled individually.

Fig. (**4.14**) shows a three-phase VSC composed of three, single-phase full-bridges. The outputs of each full-bridge are connected to a single-phase transformer. The secondary sides of the transformers are wye-connected and then connected to the neutral point. The transformers provide voltage boosting (if required), galvanic isolation, and prevent the converter switches from short-circuiting the dc link. Similar to the previous VSCs, each phase of this VSC can be controlled individually.

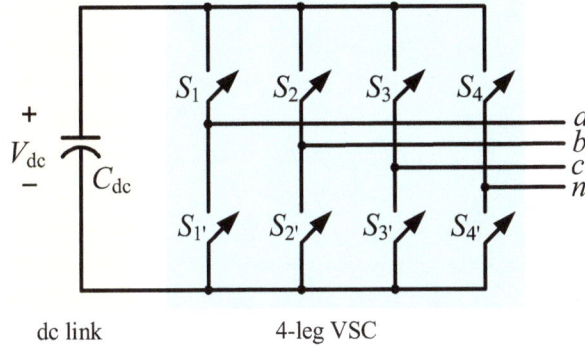

Fig. (4.13). A three-phase, 4-leg VSC.

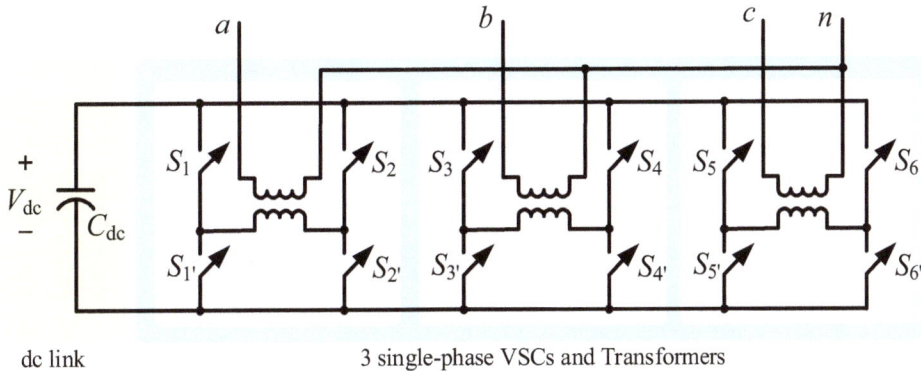

Fig. (4.14). A three-phase VSC composed of three parallel, single-phase full-bridges.

To lower the voltage stress on switching devices, reduce the *dv/dt* in voltage source, and achieve high power outputs, multilevel VSCs such as diode-clamped, flying capacitor, and cascaded H-bridge are well-known configurations that can be used in DSTATCOM structure.

Fig. (**4.15**) shows a three-phase, five-level, diode-clamped VSC. This VSC consists of a common dc link shared among 4 capacitors equally. Diodes D_1, D_2, D_3, $D_{1'}$, $D_{2'}$, and $D_{3'}$ limit the voltage stress on each switch. The output phase voltage is generated by switching each leg in eight intervals, which is shown for phase-a in Table **4.1**. A group of four switches are conducting (ON) for every interval. In the first interval, the switches from S_1 to S_4 are switched and the resultant voltage level is increased to $V_{an} = V_{dc}/2$ where the left intervals generate the other voltage levels where they synthesizes a staircase output voltage [13].

The voltages of the other phases are generated in a similar way but with a phase shift of $\pm 120°$.

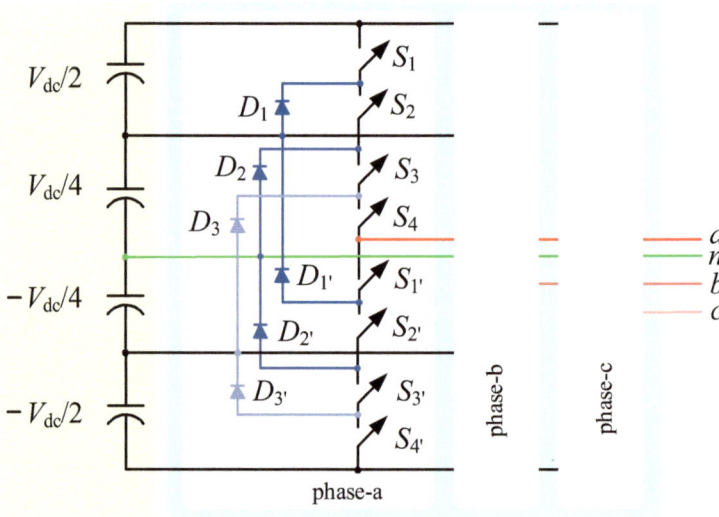

Fig. (4.15). A three-phase, five-level, diode-clamped VSC.

Table 4.1. Output voltages and switching sequences of a three-phase, five-level, diode-clamped-VSC.

Phase Voltage	Switching Sequence							
	S_1	S_2	S_3	S_4	$S_{1'}$	$S_{2'}$	$S_{3'}$	$S_{4'}$
$V_{dc}/2$	1	1	1	1	0	0	0	0
$V_{dc}/4$	0	1	1	1	1	0	0	0
0	0	0	1	1	1	1	0	0
$-V_{dc}/4$	0	0	0	1	1	1	1	0
$-V_{dc}/2$	0	0	0	0	1	1	1	1

Fig. (**4.16**) shows a three-phase, five-level, flying capacitor VSC. In this scheme, the dc link capacitors are supported by auxiliary capacitors that replace the diodes in the diode-clamped VSC. An m-level flying-capacitor VSC requires $(m-1)$ dc link capacitors along with $(m-1)(m-2)/2$ auxiliary capacitors. Although this increases the number of capacitors in the system, it results in a flexible output voltage generation. The auxiliary capacitors of C_1, C_2, and C_3 are pre-charged to $V_{dc}/4$, $V_{dc}/2$, and $3V_{dc}/4$, respectively. This VSC prevents output filter requirement; however, the increment of m can limit the precise charge control of the capacitors.

Table **4.2** shows the switching combinations for voltage generation in phase-a by this VSC [13].

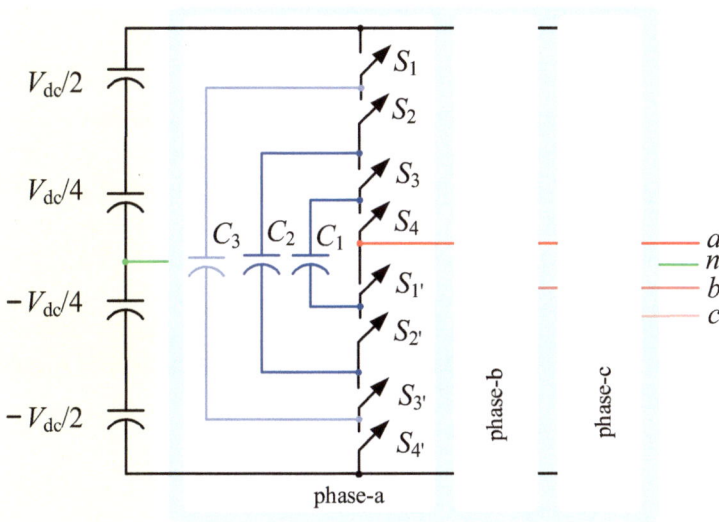

Fig. (4.16). A three-phase, five-level, flying-capacitor VSC.

Table 4.2. Output voltages and switching sequences of a three-phase, five-level, flying-capacitor VSC.

Phase Voltage	Switching Sequence							
	S_1	S_2	S_3	S_4	$S_{1'}$	$S_{2'}$	$S_{3'}$	$S_{4'}$
$V_{dc}/2$	1	1	1	1	0	0	0	0
$V_{dc}/4$	1	1	1	0	1	0	0	0
	0	1	1	1	0	0	0	1
	1	0	1	1	0	0	1	0
0	1	1	0	0	1	1	0	0
	0	0	1	1	0	0	1	1
	1	0	1	0	1	0	1	0
	1	0	0	1	0	1	1	0
	0	1	0	1	0	1	0	1
	0	1	1	0	1	0	0	1

(Table 4.2) contd.....

Phase Voltage	Switching Sequence							
	S_1	S_2	S_3	S_4	$S_{1'}$	$S_{2'}$	$S_{3'}$	$S_{4'}$
	1	0	0	0	1	1	1	0
$-V_{dc}/4$	0	0	0	1	0	1	1	1
	0	0	1	0	1	0	1	1
$-V_{dc}/2$	0	0	0	0	1	1	1	1

Fig. (**4.17**) shows a three-phase, five-level, cascaded VSC where each phase includes two series-connected full-bridges. As seen from this figure, each phase of the system can be composed of several series-connected full-bridges, each connected to an isolated dc link. Each full-bridge can generate three voltage levels of $\pm V_{dc}$ and 0. Compared to the diode-clamped and flying-capacitor VSCs, cascaded VSCs have a higher switching frequency and increased power capacity which is equally shared by each module. In addition, a cascaded VSC-based DSTATCOM can eliminate harmonics and compensate reactive power better than the other multilevel VSCs. Also, different energy storages can be integrated as the system includes numerous isolated dc links.

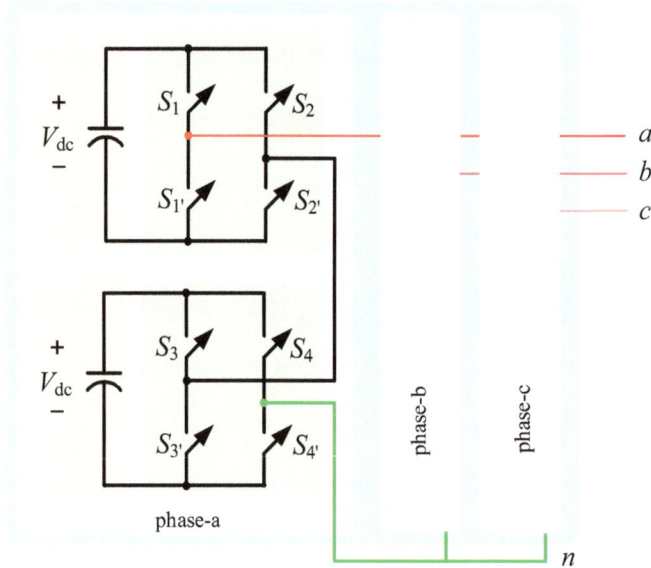

Fig. (4.17). A three-phase, five-level, cascaded VSC.

Filter Structures

A passive filter is required at the output of each phase of a VSC to eliminate the high switching frequency harmonic components in the output voltage and currents. Three common types of passive filters are the inductive filter (L), the inductive-capacitive (LC) filter, and the inductive-capacitive-inductive (LCL) filter. In all these three filters, the inductive components are usually connected to a small resistor (either in series or parallel) to increase the stability of the filter and to provide some passive damping for the filter at its resonance frequency in comparison to the pure reactive filters. These filters are briefly discussed below:

L filter, the simplest passive filter, is an inductor only. Usually, it is desired that this filter minimizes the output current ripples of the VSC below 25% while the voltage drop across that should be below 10%. The disadvantage of the L filters is that they require higher inductance to decrease the total harmonic distortion (THD). Also, the switching frequency is restricted to decrease the power losses. Fig. (**4.18**) shows a VSC with an L filter at its output where R_1 and L_1 are the equivalent resistance and inductance of the filter. The filter performance can be determined by its cut-off frequency. A larger R_1 will increase the cut-off frequency of the filter; and thereby, larger low frequency components will be removed. However, a larger L_1 decreases the cutoff frequency. On the other hand, L_1 is effective on the magnitude of medium and high frequencies while it is ineffective on the magnitude of low frequencies [27].

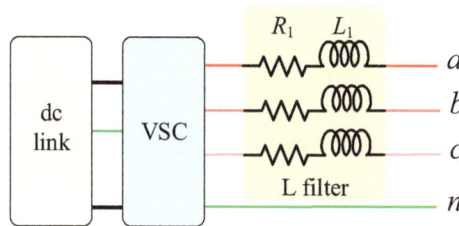

Fig. (4.18). L filter at the output of a VSC.

LC filters offer better filtering characteristics in comparison with L filters. Fig. (**4.19**) shows a VSC with an LC filter at its output. The output filter includes L_1, R_1, C and R_d that can easily eliminate the high order switching harmonics of the output current and voltage of the VSC. The transfer function of LC filter at no-load condition is

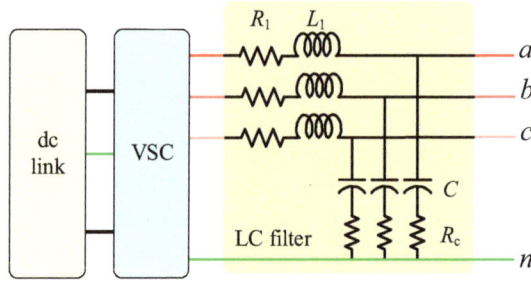

Fig. (4.19). LC filter at the output of a VSC.

$$G(s) = \frac{V_o(s)}{V_{\mathrm{VSC}}(s)} = \frac{sR_dC+1}{s^2L_1C + s(R_1+R_d)C+1} \qquad (4.31)$$

Larger L_1 reduces the cut-off frequency of the filter and eliminates the high order harmonics in the output currents of the VSC while larger C increases the cut-off frequency. However, larger L_1 leads to a larger size of the filter, higher costs, and slower dynamic response [27].

The important characteristic of LCL filters is their large gain at low frequencies and harmonic elimination at high frequencies. The harmonic elimination ability of LCL filters is better than L filters. In addition, the filtering of high frequency harmonics is limited in L filters while LCL filters can address them properly. Fig. (**4.20**) shows a VSC with an LCL filter at its output. Neglecting resistors R_1 and R_2, the transfer function of the LCL filter at no-load condition can be expressed as

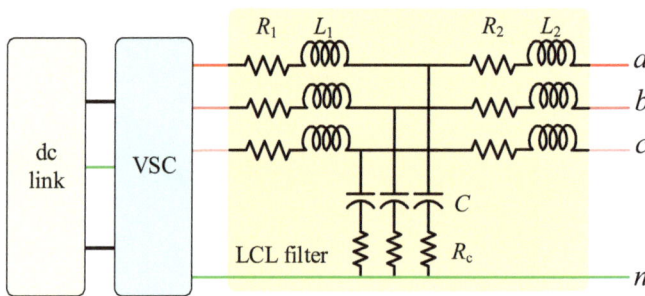

Fig. (4.20). LCL filter at the output of a VSC.

$$G(s) = \frac{V_o(s)}{V_{\mathrm{VSC}}(s)} = \frac{sR_dC+1}{s^3L_1L_2C + s^2(L_1+L_2)R_cC + s(L_1+L_2)} \qquad (4.32)$$

Variations of L_1, R_c, and C have the same impact on the LCL filter as they had on the LC filter. L_2 affects the gain in the low and medium frequencies while it has no effect on high frequency attenuation. The resonance peak of LCL filter requires special attention in design to obtain proper control of VSC and handling the utility grid disturbances. Hence, active damping techniques are often included in the VSC controller [27].

For DSTATCOM applications, usually LC and LCL filters are preferred. Fig. (**4.21**) shows the single-phase equivalent circuit of a VSC, with its LC and LCL filters. The voltage across the capacitor is denoted by v_c. Resistances R_1 and R_2 are associated with the inductances L_1 and L_2 respectively, arising due to their finite quality factor or represent a deliberately series-connected small resistor for damping issues. Also, note that if a transformer is used in the VSC structure, such as the configuration in Fig. (**4.14**), the equivalent resistance and inductance of the transformer can be merged within R_1 and L_1. For a DSTATCOM that is used for voltage regulation, a VSC with an LC filter and operating in voltage control mode is preferred while for a DSTATCOM that is used for load compensation, a VSC with an LCL filter and operating in current control mode is more suitable.

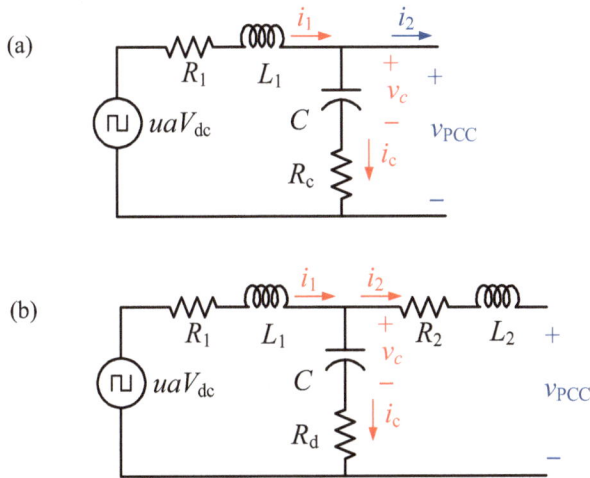

Fig. (4.21). Equivalent single-phase representation of a VSC with an LC and LCL filter.

Modulation in VSCs

To generate the desired voltage or current waveform at the output of a VSC and filter system, each switch in the VSC structure has a special driver to turn on and off based on the digital signal sent by the controller. Thus, an important task of the control system of a DSTATCOM lies in the appropriate turning on and off of the switches of the VSC. The most famous closed-loop control techniques are proportional-integral (PI) or state-feedback which provide the controller with the capability to monitor the actual generated current or voltage waveforms and update the reference signals appropriately. These signals are then usually applied to the switches through PWM-based techniques such as sinusoidal PWM (SPWM), space vector, and selective harmonic elimination, or non-linear techniques such as the hysteresis. The selected technique can affect the efficiency of a VSC considering the level of the switching losses and induced harmonics.

SPWM is one of the widely used modulation techniques in VSCs as its implementation is simple. The SPWM technique is based on comparing the reference (desired) output current or voltage of the VSC with a triangular carrier waveform (with a switching frequency of f_s) that is arranged according to required switching outputs. The ratio of the amplitude of the reference signal (also referred to as the modulating waveform) to the carrier waveform is known as modulation index (M_i). When $0 < M_i \leq 1$, the output voltage of a VSC is $\sqrt{3}V_{dc}M_i/2$. Thus, an important point of SPWM technique is that a distortion in the dc voltage will cause an amplitude distortion in the output current or voltage waveform of the VSC [27].

Selective harmonic elimination technique aims to turn on/off the switches such that some predefined harmonic orders are eliminated by defining the exact switching angles. This is achieved through the solution of a Fourier series analysis which will be then modeled as a lookup table for the controller. Thus, the complexity of the analytical calculation of the harmonic orders to be eliminated is the main disadvantage of this technique [27].

Space vector technique is a widely used method in multilevel VSCs as it can lead to a 15% higher output voltage generation with respect to SPWM. The switching signals of a space vector technique are generated by using the switching vector as a point in complex space of (α, β) of Fig. (**4.22**) [27]. This figure shows some vectors that are categorized into small vectors of V_1 to V_6, middle vectors of V_8, V_{10}, V_{12}, V_{14}, V_{16}, V_{18}, and large vectors of V_7, V_9, V_{11}, V_{13}, V_{15}, and V_{17}. Table **4.3** shows the space vector PWM switching states for a three-level VSC [27]. The vector definition and a look-up table requirement make this method complicated comparing to SPWM.

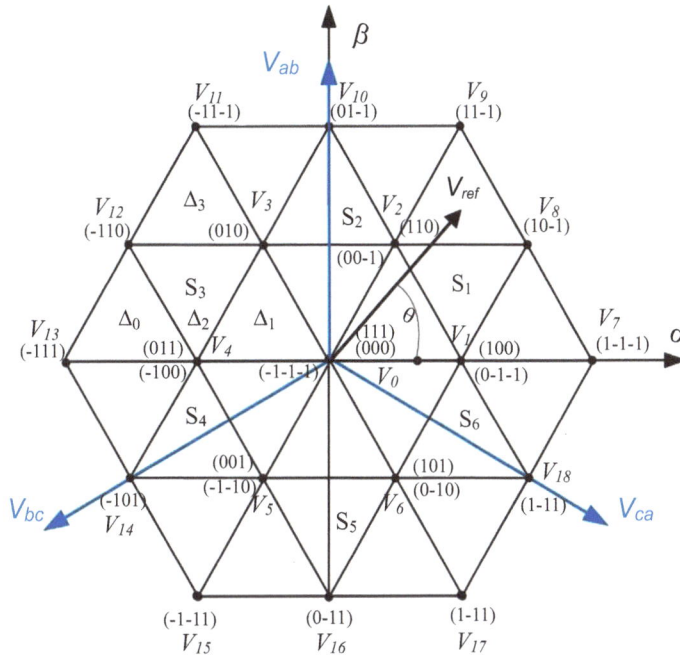

Fig. (4.22). Space vector modulation representation for a three-level VSC.

Table 4.3. Switching states for a three-level VSC using the space vector technique.

Phase Voltage	Switching Symbol	S_1	S_2	$S_{1'}$	$S_{2'}$	D_1	D_2
$V_{dc}/2$	1	1	1	0	0	0	0
0	0	0	1	1	0	0 or 1 depending on voltage polarity	
$-V_{dc}/2$	-1	0	0	1	1	0	0

The hysteresis controller aims to keep the output current or voltage waveform in the desired hysteresis band but it is allowed to oscillate between the upper and lower bands of the hysteresis. In this technique, the turn on and off of a switch varies through a cycle and it depends on the error between the reference signal and the actual generated signal. Thus, this technique does not impose a fixed switching frequency for the switches like those in SPWM.

Current-Controlled DSTATCOM

PI controllers are the most widely used control technique to regulate the output current or voltage waveform of a VSC and filter system to the reference (desired) value. In addition to them, other techniques such as fuzzy control, nonlinear-control and modern (state-feedback) control are also used. Among different modern control techniques, linear quadratic regulator (LQR) is a suitable one which has demonstrated its effectiveness in numerous applications [11]. This technique is discussed in details below:

Consider again the single-phase equivalent circuit of Fig. (**4.21**). In this figure, uaV_{dc} represents the VSC output voltage (considering a transformer with a turns ratio of 1:*a*) where *u* is the switching function of the switches. For simplicity, consider a 2-level (bipolar) switching; hence, *u* takes only ±1, and a voltage of $\pm V_{dc}$ is created at the output of the full-bridge. Note that uaV_{dc} is a square waveform with a positive peak of $+aV_{dc}$ and a negative peak of $-aV_{dc}$. If a 3-level (unipolar) switching is considered, *u* will take 0 in addition to ±1 and three voltage levels of 0 and $\pm V_{dc}$ will be created at the output of the full-bridge [15]. The main function of the VSC switching controller is to generate *u*. As discussed before, each phase of the VSC can be controlled independently. Thus, in below, the discussions are for one phase only, which will be duplicated for the other phases.

First, let us consider a current-controlled DSTATCOM with an LCL filter at the output of the VSC (Fig. **4.21b**). The dynamic behavior of the VSC and filter can be given by differential equations of

$$
\begin{aligned}
i_1 &= i_c + i_2 \\
uaV_{dc} &= R_1 i_1 + L_f \, di_1/dt + v_c + R_c i_c \\
v_c &= L_2 \, di_2/dt + v_{PCC} - R_c i_c
\end{aligned}
\tag{4.33}
$$

where i_1 is the current passing through inductor L_1, i_2 is the current passing through inductor L_2 and i_c is the current flowing into capacitor C.

Now, let the state vector (**x**) for the system of VSC and filter be

$$
\mathbf{x} = [v_c \quad i_2 \quad i_c]^T
\tag{4.34}
$$

where T is the transpose operator. The open-loop system of the VSC and filter equivalent circuit in (33) is represented in state-space domain as

$$
\begin{aligned}
\dot{x}(t) &= \mathbf{A}\,x(t) + \mathbf{B}_1\,u_c(t) + \mathbf{B}_2\,v_{PCC}(t) \\
y(t) &= \mathbf{C}\,x(t)
\end{aligned}
\tag{4.35}
$$

where

$$\mathbf{A} = \begin{bmatrix} 0 & 0 & 1/C \\ 1/L_2 & 0 & 0 \\ -(1/L_2 + 1/L_1) & -R_1/L_1 & -R_1/L_1 \end{bmatrix}$$

$$\mathbf{B}_1 = aV_{\mathrm{dc}}/L_1 \begin{bmatrix} 0 \\ 0 \\ 1 \end{bmatrix}, \quad \mathbf{B}_2 = 1/L_2 \begin{bmatrix} 0 \\ -1 \\ 1 \end{bmatrix}, \quad \mathbf{C} = \begin{bmatrix} 0 & 1 & 0 \end{bmatrix}$$

are system matrices, and u_c is the continuous-time version of u.

As in reality, these parameters are measured by sensors and then passed to the microcontroller, digital signal processor (DSP), or field-programmable gate array (FPGA) through analogue-digital converter (ADC) which discretizes the signals by the help of a sampler with sampling periods of T_s, the rest of the analysis has to be carried out in discrete-time. Eq. (35) is represented in discrete-time domain as [28]

$$\begin{aligned} \mathbf{x}[h+1] &= \mathbf{F}\mathbf{x}[h] + \mathbf{G}_1 u_c[h] + \mathbf{G}_2 v_{\mathrm{PCC}}[h] \\ y[h] &= \mathbf{C}\mathbf{x}[h] \end{aligned}$$ 　　　　(4.36)

where $\mathbf{F} = e^{\mathbf{A}.T_s}$, $\mathbf{G}_1 = \int_0^{T_s} e^{\mathbf{A}t}\mathbf{B}_1\, dt$, $\mathbf{G}_2 = \int_0^{T_s} e^{\mathbf{A}t}\mathbf{B}_2\, dt$, and h is the discrete sample index. The output of this system (y) is the output current of the DSTATCOM (i_2), while the PCC voltage (v_{PCC}) is thought as its disturbance.

To control the VSC, the references for the state variables of (34) need to be defined. The references for the output current of a DSTATCOM are discussed later in this chapter. Assuming that these references are known, and by monitoring the voltage at the PCC of the DSTATCOM, the reference for the capacitor voltage can be calculated by Kirchhoff's voltage law (KVL) as

$$v_{\mathrm{c}}^{\mathrm{ref}} = v_{\mathrm{PCC}} + i_2^{\mathrm{ref}}(R_2 + j\omega L_2)$$ 　　　　(4.37)

Let us assume, it is in the form of

$$v_{\mathrm{c}}^{\mathrm{ref}}(t) = \sqrt{2}\,V_{\mathrm{c}}^{\mathrm{ref}}\sin(\omega t + \delta_{\mathrm{c}})$$ 　　　　(4.38)

Once the reference for v_c (t) is defined, the reference for the current passing through capacitor C is calculated as

$$i_c^{\text{ref}}(t) = \sqrt{2}\,\omega C V_c^{\text{ref}} \sin(\omega t + \delta_c + 90°) \tag{4.39}$$

These references are time variant; hence, traditional PI controllers cannot be used for controlling the tracking error. Thereby, an infinite time discrete LQR-based control can be utilized which is suitable for time variant references. It is to be noted that a stochastic observer is not required if these signals are thought to be noise free [15].

Assume the tracking error (ξ) defined as

$$\xi[h] = \mathbf{x}[h] - \mathbf{x}^{\text{ref}}[h] \tag{4.40}$$

where $\mathbf{x}^{\text{ref}}$ is the reference matrix for the state vector $\mathbf{x}$. The block diagram of the feedback control system is shown in Fig. (**4.23**). As shown in this figure, $u_c[h]$ is computed from a state feedback control as

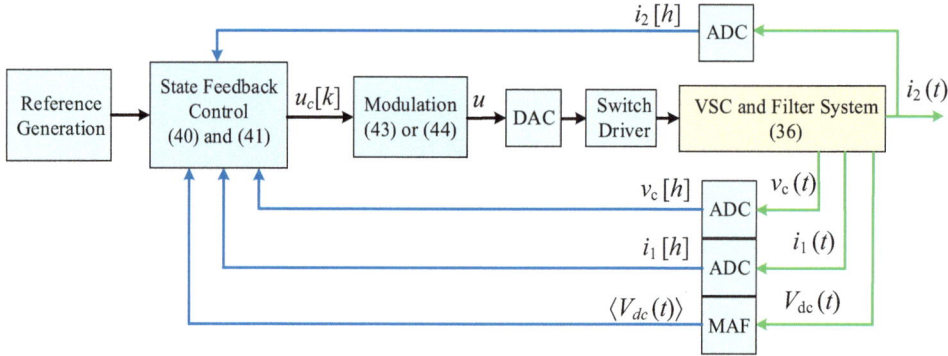

Fig. (4.23).   Closed-loop control system of a current-controlled DSTATCOM.

$$u_c[h] = -\mathbf{K}\,\xi[h] \tag{4.41}$$

where $\mathbf{K} = [k_1 k_2 k_3]$. To define $\mathbf{K}$, objective function J, chosen as [29]

$$J[h] = \sum_{h=0}^{\infty} \xi^T[h]\,\mathbf{\Psi}\,\xi[h] + u_c^T[h]\,\Theta\,u_c[h] \tag{4.42}$$

is minimized by solving the Riccati equations to obtain optimal value while satisfying system constraints in (36). In (42), Θ is a scalar control cost matrix between 0 and 1, $\mathbf{\Psi}$ is a diagonal matrix where each of its diagonal arrays represents the importance weighting for the corresponding state in $\mathbf{x}$, and $J(\infty)$ represents the objective function for the system at infinite time (steady-state). For smaller values of Θ, a higher control effort is required. It is to be noted that defining $\mathbf{K}$ takes place only once and does not need to be carried out in each sampling time because it only depends on the fixed system parameters of R_1, L_1, C, R_c, L_2, a and V_{dc}. The LQR method ensures the desired system performance provided that the variations of system load and source parameters are within acceptable limits [11].

After defining $\mathbf{K}$, $u_c[h]$ is calculated from (41). To achieve a minimum reference tracking error, u can be generated by different techniques; two of which discussed briefly below:

• Hysteresis Technique: u can be generated through a hysteresis function, denoted by hyst(.), over $u_c[h]$ within a very small bandwidth of b (*e.g.* $b = 10^{-4}$) as

$$u = \mathrm{hyst}\big(u_c[h], b\big) = \begin{cases} +1 & u_c[h] > +b \\ -1 & u_c[h] < -b \end{cases} \qquad \textbf{(4.43)}$$

This is illustrated schematically in Fig. (**4.24**).

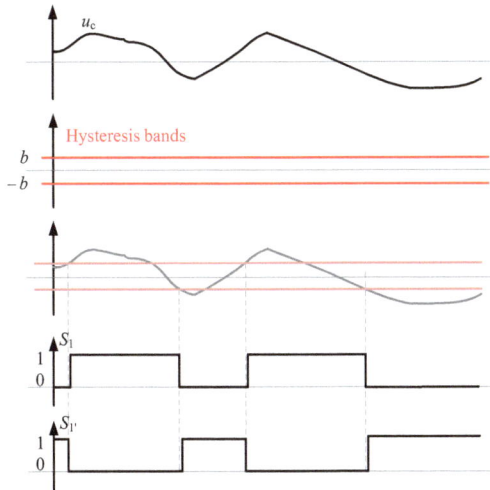

Fig. (4.24). Schematic illustration of turning on and off the VSC switches based on u_c and using the hysteresis technique.

• PWM Technique: u can be generated through a PWM function, denoted by PWM(.), which compares $u_c(h)$ with a triangular carrier waveform (v_{tri}) that varies from -1 to $+1$ with a duty ratio of 50% as

$$u = \mathrm{PWM}\left(u_c^{dis}[h], v_{tri}[h]\right) = \begin{cases} +1 & u_c[h] > v_{tri}[h] \\ -1 & u_c[h] < v_{tri}[h] \end{cases} \qquad (4.44)$$

where $u_c^{dis}[h]$ is the discretized version of with the sampling of the PWM switching frequency. This is illustrated schematically in Fig. (**4.25**).

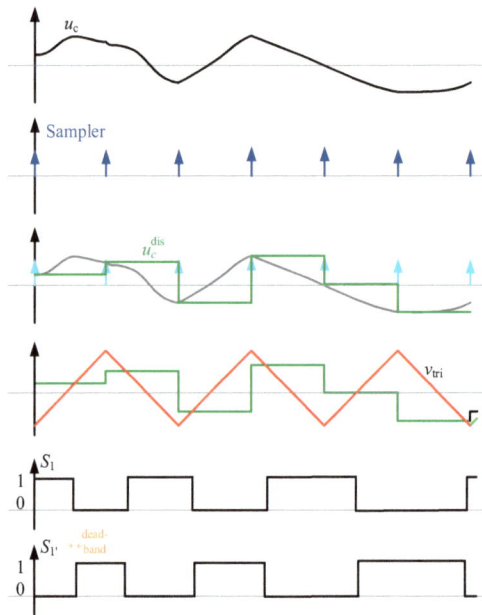

Fig. (4.25). Schematic illustration of turning on and off the VSC switches based on u_c and using the PWM technique.

u is then applied to the IGBTs or MOSFETs in the VSC through proper digital-analogue converter (DAC) and IGBT or MOSFET driver integrated circuits (ICs). Note that the top and bottom switches in each leg of the VSC operate complementarily with a suitable dead-band control to prevent short-circuiting the dc link of the DSTATCOM, as illustrated schematically in Figs. (**4.24** and **4.25**).

The power losses in IGBTs and MOSFETs depend on the switching frequency while the switching frequency depends on Θ, as well as b. Although reducing Θ and b to very small values improves the reference tracking, it might lead to high switching frequencies, and hence, high power losses. Therefore, these parameters are to be defined such that a proper reference tracking is achieved while the switching frequency and power losses in the VSC are acceptable [30].

Note that, even though a converter with a switching control represents a non-linear characteristic, this controller is designed only for the linear parts of the equations that represent the VSC and filter system, and the nonlinear parts of the equations are ignored when designing the controller [15]. The abovementioned switching control is applied for each phase independently, and the operation of the switches in each leg of the VSC is independent from the operation of the switches in the other two legs [30].

Voltage-Controlled DSTATCOM

Now, let us consider a voltage-controlled DSTATCOM with an LC filter at the output of the VSC (Fig. **4.21a**). Also, let us ignore R_c for simplification of the controller design; however, the below discussions are still applicable to the controller of DSTATCOM, if a reasonably small R_c exists in practice. The dynamic behavior of the VSC and filter can be given by differential equations of

$$ua V_{dc} = R_1 i_1 + L_f \, di_1/dt + v_c$$
$$i_1 = C \, dv_c/dt + i_2 \tag{4.45}$$

in which i_2 is the output current of the DSTATCOM (see Fig. **4.21a**).

Now, let the state vector (**x**) for this system be

$$\mathbf{x} = [v_c \quad i_1]^T \tag{4.46}$$

This open-loop system of (45) can be represented in state-space domain as

$$\dot{\mathbf{x}}(t) = \mathbf{A}\,\mathbf{x}(t) + \mathbf{B}_1\,u_c(t) + \mathbf{B}_2\,i_2(t)$$
$$y(t) = \mathbf{C}\,\mathbf{x}(t) \tag{447}$$

where

$$\mathbf{A} = \begin{bmatrix} 0 & 1/C \\ -1/L_1 & -R_1/L_1 \end{bmatrix}, \quad \mathbf{B}_1 = aV_{dc}/L_1 \begin{bmatrix} 0 \\ 1 \end{bmatrix}, \quad \mathbf{B}_2 = 1/C \begin{bmatrix} -1 \\ 0 \end{bmatrix}, \quad \mathbf{C} = \begin{bmatrix} 1 & 0 \end{bmatrix}$$

are system matrices. The output of this system (y) is the PCC voltage (vPCC) while the output current (i_2) is thought as its disturbance. As defining the reference for i1 is difficult and considering the fact that i1 is desired to have low frequencies only, instead of using i1 as a control parameter, its high frequency components (ĩ1) can be used in the control system. ĩ1 can be described and expanded in Laplace domain as [16]

$$\tilde{i}_1(s) = \frac{s}{s+\alpha} i_1(s) = \left(1 - \frac{\alpha}{s+\alpha}\right) i_1(s) = i_1(s) - \hat{i}_1(s) \tag{4.48}$$

where α is the cut-off frequency of this high-pass filter (*e.g.* α = 500 Hz) while î1 is the low frequency components of i1 and is given by

$$\hat{i}_1(s) = \frac{\alpha}{s+\alpha} i_1(s) \tag{4.49}$$

Eq. (49) can be expressed in differential equation form as

$$\frac{d\hat{i}_1}{dt} = \alpha(i_1 - \hat{i}_1) \tag{4.50}$$

Now, let us define a new state vector for the system, which includes î1, as

$$\mathbf{x}'(t) = [v_c \quad i_1 \quad \hat{i}_1]^T \tag{4.51}$$

In this case, the system of (45) can be represented with the new state-space equation of

$$\dot{\mathbf{x}}'(t) = \mathbf{A}'\mathbf{x}'(t) + \mathbf{B}'_1 u_c(t) + \mathbf{B}'_2 i_2(t)$$
$$y(t) = \mathbf{C}'\mathbf{x}'(t) \tag{4.52}$$

where

$$\mathbf{A}' = \begin{bmatrix} 0 & 1/C & 0 \\ -1/L_1 & -R_1/L_1 & 0 \\ 0 & \alpha & -\alpha \end{bmatrix}, \quad \mathbf{B}'_1 = aV_{dc}/L_1 \begin{bmatrix} 0 \\ 1 \\ 0 \end{bmatrix}, \quad \mathbf{B}'_2 = 1/C \begin{bmatrix} -1 \\ 0 \\ 0 \end{bmatrix}, \quad \mathbf{C}' = \begin{bmatrix} 1 & 0 & 0 \end{bmatrix}$$

Eq. (52) is then represented in discrete-time domain. The rest of the analysis is the same as discussed for the current-controlled DSTATCOM.

The references for the output voltages of a DSTATCOM are discussed in the rest of this chapter. The reference value for $\tilde{i}_1$ is always set to zero to minimize the high frequency components of the current flowing through L_1. Therefore, i_1 must only contain low frequency components (*i.e.* $i_1 = \hat{i}_1$). Now, the reference vector, $\mathbf{x}'^{ref}$ can be defined as

$$\mathbf{x}'^{ref} = [v_c^{ref} \quad \hat{i}_1 \quad \hat{i}_1]^T \tag{4.53}$$

The closed-loop control system of such a DSTATCOM is illustrated in Fig. (**4.26**).

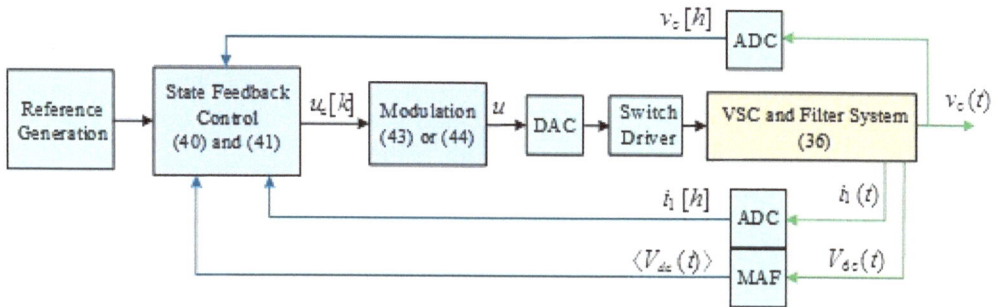

Fig. (4.26). Closed-loop control system of a voltage-controlled DSTATCOM.

Other Control Aspects of DSTATCOMs

The dc link of the DSTATCOM is usually composed of one or two series-connected dc capacitors, as shown in Fig. (**4.11** and **4.12**). The dc link voltage needs to be maintained above a certain pre-specified value for proper operation of the DSTATCOM (*i.e.*, proper tracking of the desired output voltage and current waveforms). If this voltage is below a certain limit, the tracking of the voltage and current becomes unacceptable or can even lead to the system collapse [11]. In some cases, an energy storage system such as a battery or another type of renewable energy source can be coupled to this dc link. This can help the DSTATCOM to inject active power, if required. Alternatively, the dc link can be supplied through a single-phase/three-phase connection to the same feeder in which the DSTATCOM is installed via a rectifier.

Ideally, the DSTATCOM should not inject active power. Thereby, as soon as the dc link capacitors are charged, they should uphold the charge so that the voltage across them remains unchanged. However, the DSTATCOM will have some losses due to the switching losses in the VSC, and its passive filters. If a transformer is used in the DSTATCOM structure (see Fig. **4.14**), the transformer also has some losses. If a special mechanism is not in place, these losses will result in a decrease in the charge and voltage of the dc link until it collapses. Thereby, a mechanism should compensate the losses of the DSTATCOM such that the voltage across the dc link of the DSTATCOM is regulated to the desired level (V_{dc}^{ref}). This can be achieved in two different ways, depending on the control mechanism of the DSTATCOM [11, 26].

For a voltage controlled DSTATCOM, the angle of the voltage across C (referred to as δ_c) can be calculated continuously, through a PI controller that monitors the deviation of the dc link voltage from its reference, as

$$\delta = k_1(V_{dc}^{ref} - \langle V_{dc} \rangle) + k_2 \int V_{dc}^{ref} - \langle V_{dc} \rangle \, dt \qquad (4.54)$$

where $\langle V_{dc} \rangle$ represents the averaged V_{dc} using a low pass filter or a moving averaging filter while k_1 and k_2 are the gains of PI controller. This angle variation will allow the DSTATCOM to absorb some active power which will be equal to its losses; and thereby, will regulate the dc link voltage.

For a current-controlled DSTATCOM, the output reference currents of the DSTATCOM should consider the power loss of the DSTATCOM (P_{loss}), calculated from

$$P_{loss} = k_1(V_{dc}^{ref} - \langle V_{dc} \rangle) + k_2 \int V_{dc}^{ref} - \langle V_{dc} \rangle \, dt \qquad (4.55)$$

and then incorporated in the output current references. This is shown later in this chapter.

Another very important control aspect of the DSTATCOMs in microgrids is the frequency of the reference current or voltage signal. In a traditional distribution network, frequency is approximately 50 Hz. Thus, the nominal frequency of the system can be directly used in the reference signal generation, without actually measuring the system frequency. This will not cause a vivid impact on the performance of the DSTATCOM as it will latch itself with the system frequency almost immediately. However, this is not true for microgrids as its frequency may vary largely depending on the loading conditions of the system, as discussed earlier, and it may not be as stable and steady as that in traditional distribution networks. First, assume a voltage-controlled DSTATCOM with an LC filter in which the microgrid frequency is not 50 Hz (*i.e.*, $f \neq 50$ or $\omega \neq 100\pi$) while the DSTATCOM reference voltages are synthesized with $\omega_0 = 100\pi$. Thus, the reference for the voltage waveform of phase-a of the DSTATCOM is

$$v_{c,a}^{\text{ref}}(t) = \sqrt{2}V_c^{\text{ref}} \sin(\omega_0 t + \delta_c) \tag{4.56}$$

Under such a scenario, angle δ_c calculated from the PI controller in (54) does not stabilize at a constant value and starts to vary over time. It decreases gradually if the microgrid frequency is below 50 Hz and rises continuously when the frequency is above 50 Hz. In practice, the DSTATCOM cannot operate at 50 Hz and latches to the microgrid frequency. Thus, the actual voltage that is produced at the output of the VSC and filter system becomes

$$v_{c,a}^{\text{ref}}(t) = \sqrt{2}V_c^{\text{ref}} \sin(\omega_0 t + \delta_c^{\text{new}} + \Delta\omega t) \tag{4.57}$$

where $\Delta\omega$ is the mismatch between the microgrid's actual frequency and 50 Hz (*i.e.* $\Delta\omega < 0$ when the microgrid frequency is below 50 Hz, and vice versa) and δ_1 is a constant value that regulates the power flow. From (56) and (57), one gets

$$\delta_c = \Delta\omega t + \delta_c^{\text{new}} \tag{4.58}$$

from which it can be seen that the angle drops or rise by 0.2π rad each second for 0.1 Hz variation from 50 Hz, due to the integrator component in the PI controller of (2). Also, it may reach the integrator limits and saturate, which can cause a system collapse. Correspondingly, the voltage of the dc link deviates from the desired level. If δ_c does not stabilize, the voltage across the dc link will vary. This variation can be expressed as

$$\Delta V_{\text{dc}} = 1 - \frac{\Delta\omega}{k_2 V_{\text{dc}}^{\text{ref}}} \tag{4.59}$$

As the dc link voltage reduces when the microgrid frequency is less than 50 Hz, the VSC tracking may degrade or collapse. Likewise, when the microgrid frequency is above 50 Hz, the dc link voltage increases which may damage the capacitors and the switches of the VSC.

As an example, Fig. (**4.27**) shows the continuous decrease and increase of δ_c for a DSTATCOM which is operating in a microgrid with 49.9 and 50.1 Hz while its reference voltages are produced at 50 Hz. Correspondingly, the voltage of the dc link becomes smaller or larger than the desired level.

To alleviate such problems, the current or voltage reference signals of the DSTATCOM must be produced at the microgrid frequency. To this end, the frequency of the microgrid should be calculated using a proper technique and then used in the references. Thus, the reference signals of (56) will be modified as

$$v_{c,a}^{\text{ref}}(t) = \sqrt{2}V_{cf}^{\text{ref}} \sin(2\pi\hat{f}t + \delta_c) \tag{4.60}$$

where $\hat{f}$ is the microgrid's calculated frequency.

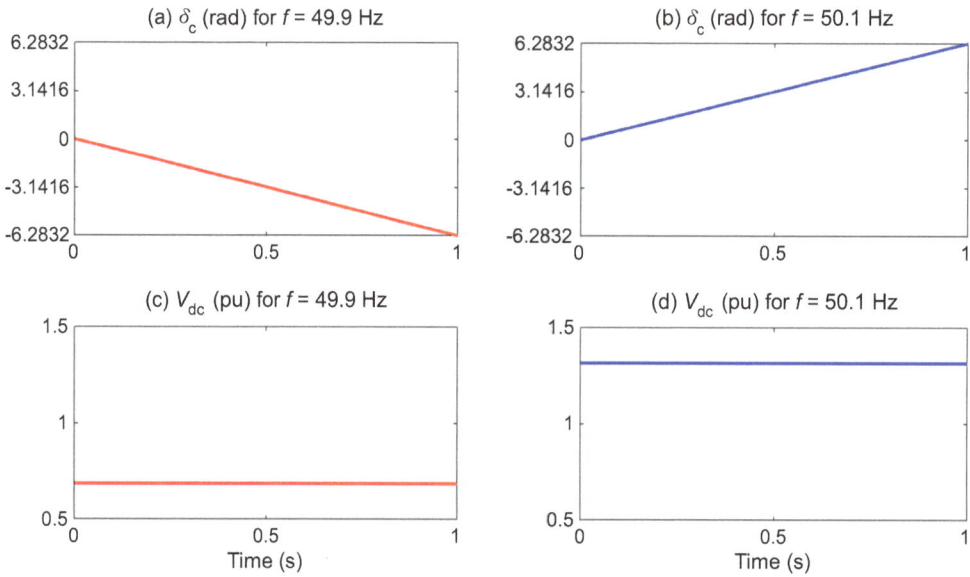

Fig. (4.27). Variation of δ_c and V_{dc} when the DSTATCOM references are generated at 50 Hz while the microgrid frequency is not 50 Hz.

APPLICATIONS OF DSTATCOM IN MICROGRIDS

This section introduces some of the applications of DSTATCOMs in microgrids. An example, realized in PSCAD/EMTDC, is provided for each application to demonstrate the performance of the DSTATCOM.

Load Compensation in a Microgrid

Consider Fig. (**4.28**) which illustrates a microgrid feeder that is supplying a nonlinear and unbalanced load. The DSTATCOM intends to compensate the unbalance and nonlinear current of the load. Thus, a current-controlled DSTATCOM with an LCL filter (Fig. **4.21b**) is connected to the feeder at a close point to the load. Let us assume the DSTATCOM as an ideal current source that injects current i_2 into to the PCC. When a nonlinear and unbalanced load is assumed, which may also be consuming a large reactive power (*i.e.*, poor power factor) and before the connection of the DSTATCOM, the current drawn by the load (i_{load}) is flowing in the microgrid feeder and has to be generated and supplied by the DERs of the microgrid. After the connection of the DSTATCOM, Kirchhoff's current law (KCL) at the PCC gives

$$i_{\text{MGF}} + i_2 = i_{\text{load}} \tag{4.61}$$

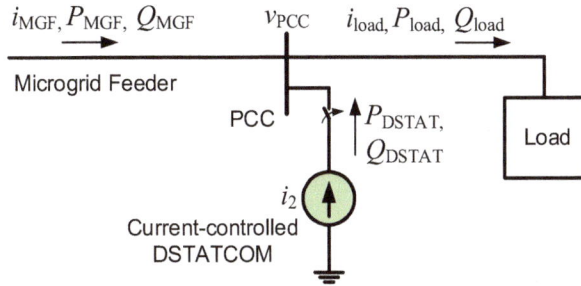

Fig. (4.28). Schematic illustration of a current-controlled DSTATCOM installed in a microgrid feeder.

It is desired that the DSTATCOM compensates the load, and thus, facilitates a balanced three-phase current drawn from the microgrid feeder into the PCC; *i.e.*,

$$i_{\text{MGF},a} + i_{\text{MGF},b} + i_{\text{MGF},c} = 0 \tag{4.62}$$

at unity power factor (all load reactive power is supplied by the DSTATCOM); *i.e.*,

$$\begin{aligned}
\angle v_{\text{PCC},a} &= \angle i_{\text{MGF},a} \\
\angle v_{\text{PCC},b} &= \angle i_{\text{MGF},b} \\
\angle v_{\text{PCC},c} &= \angle i_{\text{MGF},c}
\end{aligned} \tag{4.63}$$

However, the DSTATCOM will not be supplying any active power into the PCC (unless it is desired in a particular case which is possible if the dc link of the DSTATCOM is connected to an energy storage or dc source). Thereby, the output current of the DSTATCOM should be equal to the reactive power, harmonic, negative and zero sequence components of the load current (denoted respectively by ^{Q}, ^{H}, ^{NZ}, and ^{ZS}); *i.e.,*

$$i_2 = i_{\text{load}}^Q + i_{\text{load}}^H + i_{\text{load}}^{NS} + i_{\text{load}}^{ZS} \tag{4.64}$$

To this end, it is given in [11] that the references of the output currents of the DSTATCOM will be

$$\begin{bmatrix} i_{2,a} \\ i_{2,b} \\ i_{2,c} \end{bmatrix}^{\text{ref}} = \begin{bmatrix} i_{\text{load}a} \\ i_{\text{load}b} \\ i_{\text{load}c} \end{bmatrix} - \frac{P_{\text{load}} + P_{\text{loss}}}{v_{t,a}^2 + v_{t,b}^2 + v_{t,c}^2} \begin{bmatrix} v_{t,a} \\ v_{t,b} \\ v_{t,c} \end{bmatrix} \tag{4.65}$$

in which is the fundamental positive sequence component of v_{PCC}, the three phases are denoted by $_a$, $_b$ and $_c$, and P_{loss} is considered to regulate the dc link voltage of the DSTATCOM.

If the DSTATCOM balances the current drawn from the microgrid feeder, and assuming that its upstream was also balanced, the PCC voltages also become balanced indirectly. However, the DSTATCOM does not regulate it.

Assume that the system of (Fig. **4.28**) is at steady-state condition initially while the DSTATCOM is not connected yet. Fig. (**4.29**) illustrates the uncompensated current drawn from the feeder and the unbalanced voltages at the PCC. Fig. (**4.30**) illustrates these parameters after the DSTATCOM connection. As seen from this figure, the PCC voltage and the current drawn from the microgrid feeder become sinusoidal and balanced while the uncompensated current only flows from the PCC of the DSTATCOM towards the load [26].

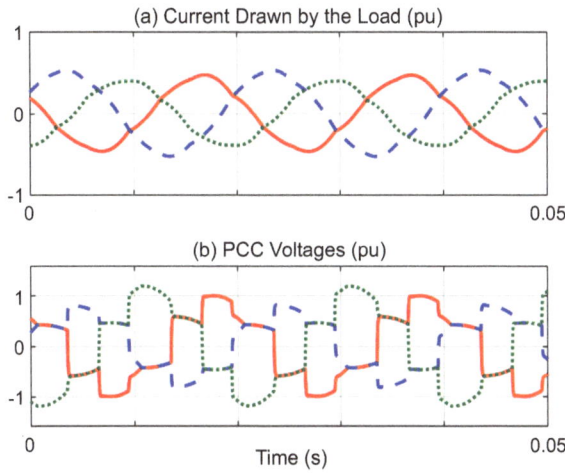

Fig. (4.29). Sample results for the system of Figure 28 before the connection of DSTATCOM.

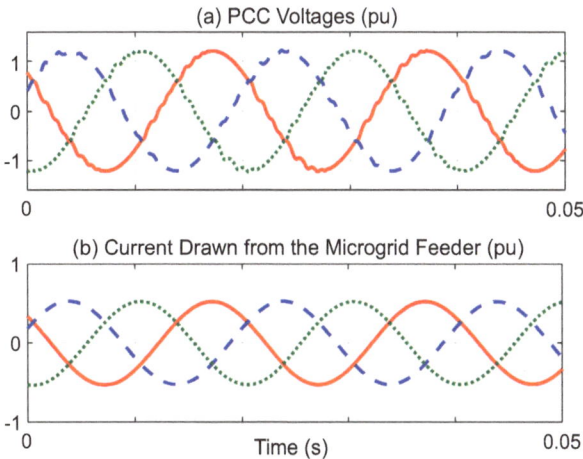

Fig. (4.30). Sample results for the system of (Fig. **4.28**) after the connection of a DSTATCOM operating in load compensation mode.

Power Factor Correction in a Microgrid

DSTATCOM can be used in microgrids for power factor correction. In this application, the DSTATCOM's principle of the operation is the same as when used in a traditional distribution network. Consider the system of (Fig. **4.31**) which illustrates a voltage-controlled DSTATCOM with an LC filter (Fig. **4.21a**), installed in a microgrid feeder supplying a load with some reactive power demand. The aim of the DSTATCOM application is to force the reactive power flow in the microgrid feeder (from the upstream side towards the PCC of the DSTATCOM) to zero, which will subsequently increase the capacity of the feeder for active power flow to the other loads of the feeder. In this application, the DSTATCOM does not regulate the voltage magnitude at the PCC of the DSTATCOM to any particular level. To facilitate unity power factor at the PCC of the DSTATCOM, the reference for the PCC voltage magnitude can be defined with respect to the reactive power flowing from the feeder upstream towards the DSTATCOM PCC (Q_{MGF}) and using a PI controller as [26]

$$V_c^{ref} = -\left(k_P Q_{MGF} + k_I \int Q_{MGF}\, dt\right) \tag{4.66}$$

where V_c^{ref} is the magnitude of the reference voltage across C and is then used to generate the reference of the voltage across C as

$$
\begin{aligned}
v_{c,a}^{ref} &= \sqrt{2} V_c^{ref} \sin(2\pi \hat{f} t + \delta_c) \\
v_{c,b}^{ref} &= \sqrt{2} V_c^{ref} \sin(2\pi \hat{f} t + \delta_c - 120°) \\
v_{c,c}^{ref} &= \sqrt{2} V_c^{ref} \sin(2\pi \hat{f} t + \delta_c + 120°)
\end{aligned}
\tag{4.67}
$$

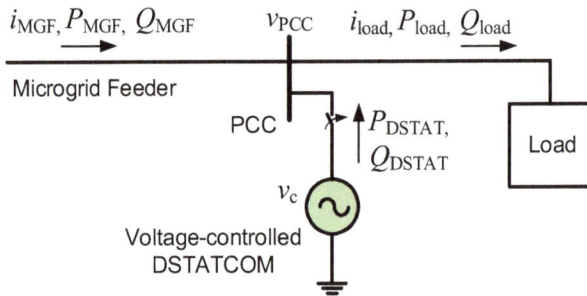

Fig. (4.31). Schematic illustration of a voltage-controlled DSTATCOM installed in a microgrid feeder.

As an example, assume the system of (Fig. **4.31**) at steady-state condition initially while the DSTATCOM is not connected yet. The DSTATCOM connects at $t = 0.2$ s. Fig. (**4.32**) illustrates the active and reactive power flowing in the feeder from the upstream towards the PCC of the DSTATCOM, as well as the injected active and reactive power by the DSTATCOM. As seen from this figure, after the DSTATCOM connection, the reactive power flow in the microgrid feeder becomes zero at the upstream side of the DSTATCOM and the DSTATCOM starts to inject the load reactive power. It is also seen that the DSTATCOM consumes a small amount of active power to compensate its losses, which results in a slight increase in the active power flow from the feeder upstream towards the DSTATCOM. The overshoot and settling time of the system can be improved by selecting the gains of the PI controller in (66) optimally. This figure also illustrates that the PCC voltage magnitude settles at a new level, as it is not directly regulated [26].

Voltage Regulation and Balancing in a Microgrid

Let us again consider (Fig. **4.31**) in which the DSTATCOM intends to regulate

the voltage magnitude at its PCC to a desired level (V_c^{ref}) such that any voltage variations in the microgrid feeder are not exposed to the load. Thus, a voltage-controlled DSTATCOM with an LC filter (Fig. **4.21a**) is utilized. The DSTATCOM can also force the three-phase voltage at its PCC to be balanced (*i.e.*, 120 degree phase shifts among the phases). The output voltage reference for the DSTATCOM will be similar to (67).

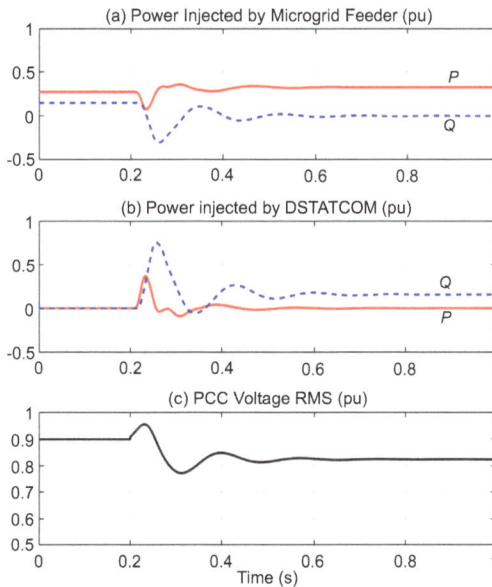

Fig. (4.32). Sample results for the system of (Fig. **4.31**), before and after the connection of a DSTATCOM operating in power factor correction mode.

To this end, the DSTATCOM exchanges some reactive power with the feeder such that the voltage magnitude is regulated to the desired level (*i.e.*, injection of the reactive power for increasing the voltage magnitude, and vice versa); however, the controller does not need to calculate the amount of this reactive power. It is to be noted that if the PCC voltage is desired to be equal to or larger than the voltage magnitude at the upstream side of the feeder, some reactive power has to flow in the feeder from the DSTATCOM towards the upstream, as discussed in [26]. However, if no reverse reactive power flow is allowed, the voltage magnitude at the PCC cannot reach the magnitude of the upstream side of the feeder. It is noteworthy that in some cases, especially in low voltage feeders or feeders with a large R/X ratio, the DSTATCOM may fail to regulate its PCC voltage to the desired level by reactive power exchange only. In those cases, the DSTATCOM can regulate the voltage if it injects active power as well. Again, the amount of the active power exchange is not required to be calculated by the controller as it will happen automatically. However, this is only applicable if the DSTATCOM is supplied by an energy source such a battery, or a renewable source with a sufficient capacity.

As an example, assume that the system of (Fig. **4.31**) is at steady-state condition initially Also, assume that the feeder voltage reduces at $t = 0.1$ s and increases at $t = 0.3$ s. Fig. (**4.33**) shows the three-phase instantaneous voltages of the PCC, as well as the current flowing in the microgrid feeder in this period. As seen from this figure, the DSTATCOM maintains its PCC voltage at the desired level irrespective of the voltage variations [26].

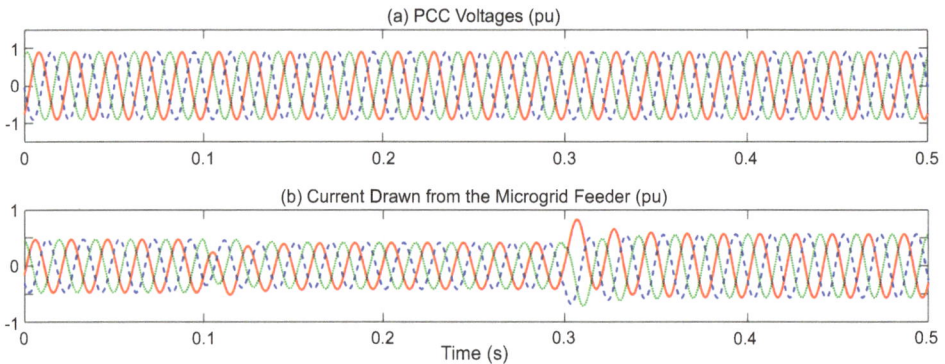

Fig. (4.33). Sample results for the system of (Fig. **4.31**), in the presence of a DSTATCOM operating in voltage regulation and balancing mode.

Combination of Power Quality Improvement and Energy Harvesting from a Renewable Source in Microgrid

Let us again consider the system of (Fig. **4.27**). The DSTATCOM intends to compensate the nearby nonlinear load. However, it is assumed that the dc link of the DSTATCOM is supplied via a renewable energy-based source. Thus, it can also inject active power and share a portion of the demand of the microgrid according to its renewable source's capacity.

Depending on the power requirement of the nonlinear load, the DSTATCOM will work in different modes [14]:

- Mode-1: If the demand of the nonlinear load is less than the capacity of the DSTATCOM, the DSTATCOM will supply the nonlinear load fully while the rest of its power will be fed into the microgrid feeder.

- Mode-2: If the demand of the nonlinear load is more than the capacity of the DSTATCOM, the extra power requirement will be supplied by the microgrid feeder.

In this application, the DSTATCOM supplies a current to the PCC that balances the PCC voltage; and therefore, a balanced and non-harmonic current will be drawn from or injected to the microgrid feeder. For this purpose, a current-controlled DSTATCOM with an LCL filter is suitable.

The references for the output current of the DSTATCOM are given by [14]

$$\begin{bmatrix} i_{2,a} \\ i_{2,b} \\ i_{2,c} \end{bmatrix}^{\text{ref}} = \begin{bmatrix} i_{\text{load}a} \\ i_{\text{load}b} \\ i_{\text{load}c} \end{bmatrix} - \frac{1}{v_{t,a}^2 + v_{t,b}^2 + v_{t,c}^2} \begin{bmatrix} 3P_{\text{MGF}}v_{t,a} + \sqrt{3}Q_{\text{MGF}}\left(v_{t,b} - v_{t,c}\right) \\ 3P_{\text{MGF}}v_{t,b} + \sqrt{3}Q_{\text{MGF}}\left(v_{t,c} - v_{t,a}\right) \\ 3P_{\text{MGF}}v_{t,c} + \sqrt{3}Q_{\text{MGF}}\left(v_{t,a} - v_{t,b}\right) \end{bmatrix} \quad \textbf{(4.68)}$$

where P_{MGF} and Q_{MGF} are respectively the active and reactive power drawn from the microgrid feeder. From (3), the references for the output current of the DSTATCOM will be derived separately for each modes.

In mode-1, it is expected P_{MGF} and Q_{MGF} to be negative (*i.e.*, the excessive power of the DSTATCOM is flowing into the microgrid feeder). So, the power that can be injected by the DSTATCOM to the PCC will be

$$\begin{aligned} P_{\text{DSTAT}}^{\text{cap}} &= P_{\text{MGF}} - \left(P_{\text{load}} + P_{\text{loss}}\right) \\ Q_{\text{DSTAT}}^{\text{cap}} &= Q_{\text{MGF}} - Q_{\text{load}} \end{aligned} \quad \textbf{(4.69)}$$

where $P_{\mathrm{DSTAT}}^{\mathrm{cap}}$ and $Q_{\mathrm{DSTAT}}^{\mathrm{cap}}$ are respectively the capacity (maximum available) active and reactive power output of the DSTATCOM's renewable source, calculated based on the maximum current that it can inject. Thereby, the references of the output current of the DSTATCOM will be

$$
\begin{bmatrix} i_{2,a} \\ i_{2,b} \\ i_{2,c} \end{bmatrix}^{\mathrm{ref}} = \begin{bmatrix} i_{\mathrm{load},a} \\ i_{\mathrm{load},b} \\ i_{\mathrm{load},c} \end{bmatrix}
$$

$$
+ \frac{1}{v_{t,a}^2 + v_{t,b}^2 + v_{t,c}^2} \begin{bmatrix} 3\left(P_{\mathrm{DSTAT}}^{\mathrm{cap}} - \left(P_{\mathrm{load}} + P_{\mathrm{loss}}\right)\right) v_{t,a} + \sqrt{3}(Q_{\mathrm{DSTAT}}^{\mathrm{cap}} - Q_{\mathrm{load}})(v_{t,b} - v_{t,c}) \\ 3\left(P_{\mathrm{DSTAT}}^{\mathrm{cap}} - \left(P_{\mathrm{load}} + P_{\mathrm{loss}}\right)\right) v_{t,b} + \sqrt{3}(Q_{\mathrm{DSTAT}}^{\mathrm{cap}} - Q_{\mathrm{load}})(v_{t,c} - v_{t,a}) \\ 3\left(P_{\mathrm{DSTAT}}^{\mathrm{cap}} - \left(P_{\mathrm{load}} + P_{\mathrm{loss}}\right)\right) v_{t,c} + \sqrt{3}(Q_{\mathrm{DSTAT}}^{\mathrm{cap}} - Q_{\mathrm{load}})(v_{t,a} - v_{t,b}) \end{bmatrix}
\quad \textbf{(4.70)}
$$

In mode-2, it is expected I_{MGF}, P_{MGF}, and Q_{MGF} to be positive (*i.e.*, the rest of the nonlinear load's demand is supplied by the microgrid feeder). So, the power that is injected from the MG feeder will be

$$
\begin{aligned}
P_{\mathrm{MGF}} &= \left(P_{\mathrm{load}} + P_{\mathrm{loss}}\right) - P_{\mathrm{DSTAT}} = \left(P_{\mathrm{load}} + P_{\mathrm{loss}}\right)(1 - \lambda_P) \\
Q_{\mathrm{MGF}} &= Q_{\mathrm{load}} - Q_{\mathrm{DSTAT}} = Q_{\mathrm{load}} - \lambda_Q Q_{\mathrm{load}} = Q_{\mathrm{load}}(1 - \lambda_Q)
\end{aligned}
\quad \textbf{(4.71)}
$$

where $0 < \lambda_P < 1$ and $0 < \lambda_Q < 1$ are respectively fractions of the active and reactive power supplied by DSTATCOM to the nonlinear load. Thereby, the output reference currents of the DSTATCOM will be

$$
\begin{bmatrix} i_{2,a} \\ i_{2,b} \\ i_{2,c} \end{bmatrix}^{\mathrm{ref}} = \begin{bmatrix} i_{\mathrm{load},a} \\ i_{\mathrm{load},b} \\ i_{\mathrm{load},c} \end{bmatrix}
$$

$$
- \frac{1}{v_{t,a}^2 + v_{t,b}^2 + v_{t,c}^2} \begin{bmatrix} 3(1 - \lambda_P)\left(P_{\mathrm{load}} + P_{\mathrm{loss}}\right) v_{t,a} + \sqrt{3}(1 - \lambda_Q)Q_{\mathrm{load}}(v_{t,b} - v_{t,c}) \\ 3(1 - \lambda_P)\left(P_{\mathrm{load}} + P_{\mathrm{loss}}\right) v_{t,b} + \sqrt{3}(1 - \lambda_Q)Q_{\mathrm{load}}(v_{t,c} - v_{t,a}) \\ 3(1 - \lambda_P)\left(P_{\mathrm{load}} + P_{\mathrm{loss}}\right) v_{t,c} + \sqrt{3}(1 - \lambda_Q)Q_{\mathrm{load}}(v_{t,a} - v_{t,b}) \end{bmatrix}
\quad \textbf{(4.72)}
$$

As an example, consider the system of Fig. (**4.27**) in which the load causes a 9.3 and 1.47% unbalance in the current drawn from the microgrid feeder and the voltage of PCC, respectively. It also induces a 9.8 and 8.2% THD in the current and voltage. The DSTATCOM intends to compensate the load unbalance and harmonics and while it is connected to a renewable energy source and thereby can supply active power. The DSTATCOM connects at $t = 0.04$ s. Fig. (**4.34**) shows the drawn current by the load, the injected current by the DSTATCOM, the current flowing in the microgrid feeder, and the PCC voltage of the DSTATCOM, before and after the DSTATCOM connection. As seen from this figure, the DSTATCOM injects the required current to its PCC to balance the voltage; and therefore, the current drawn from the microgrid feeder is forced to be balanced. Fig. (**4.35**) depicts the unbalance and THD of current drawn from the microgrid feeder, and the PCC voltage, before and after the DSTATCOM connection. The current and voltage unbalance becomes less than 0.2 and 0.05%, respectively, while their THD values become less than 0.4 and 0.2%, respectively. Fig. (**4.36**) shows the frequency spectrum and proves the reduction of the harmonic orders of the current and voltage after the DSTATCOM connection. The DSTATCOM also compensates the reactive power demand of the load and thus facilitates a unity power factor at the PCC. Fig. (**4.37**) shows the instantaneous voltage and current waveforms for phase-a where the current-voltage phase difference (*i.e.*, the low power factor) is vivid before the DSTATCOM connection while it reduces to zero (*i.e.*, unity power factor) after that [14].

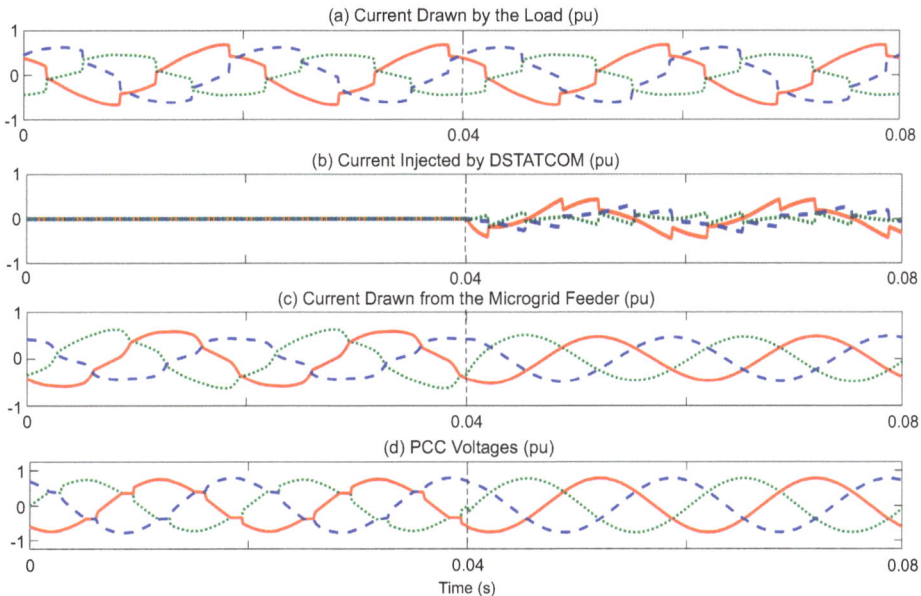

Fig. (4.34). Sample results for the system of (Fig. **4.27**), before and after the connection of a DSTATCOM operating in load compensation and energy harvesting mode.

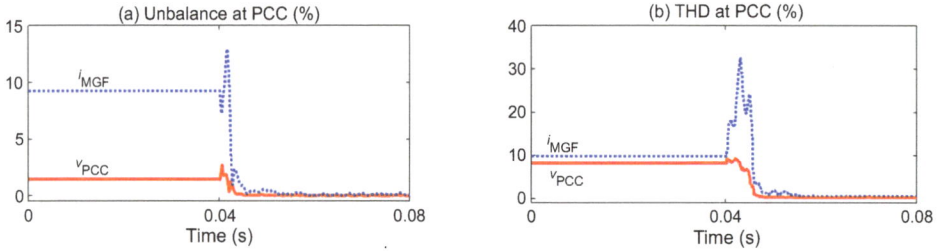

Fig. (4.35). Sample results for the system of (Fig. **4.27**), before and after the connection of a DSTATCOM operating in load compensation and energy harvesting mode.

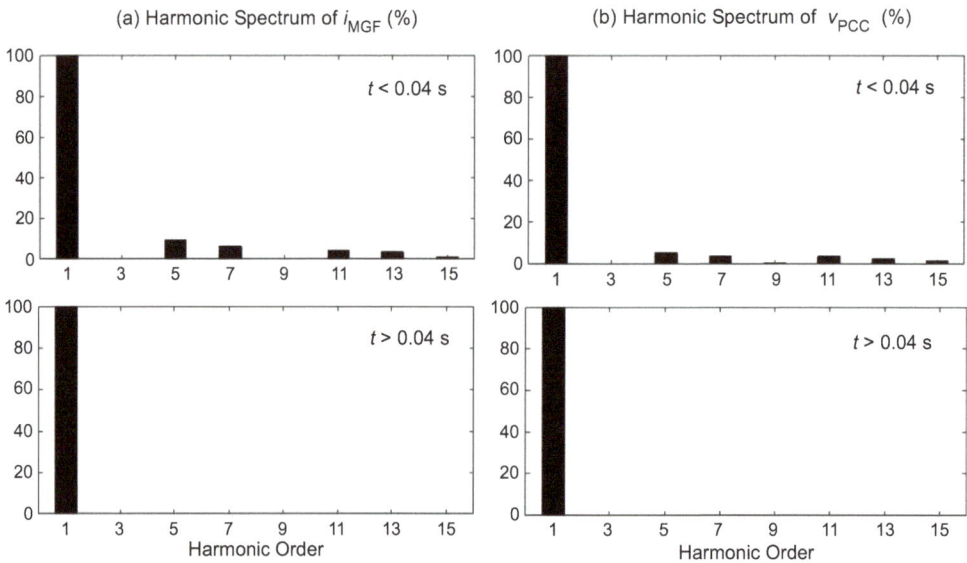

Fig. (4.36). Sample results for the system of Fig. (**4.27**), before and after the connection of a DSTATCOM operating in load compensation and energy harvesting mode.

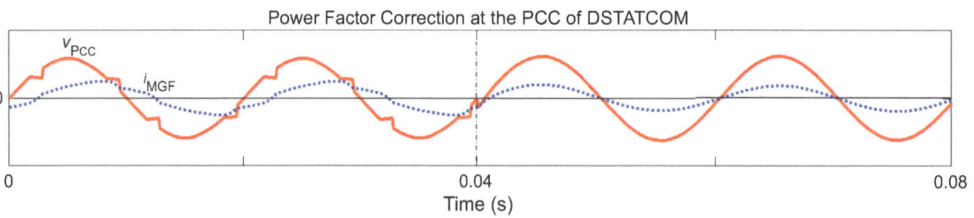

Fig. (4.37). Sample result for the system of Fig. (**4.27**), before and after the connection of a DSTATCOM operating in load compensation and energy harvesting mode.

Now, let us consider the operation of the DSTATCOM in mode-1 and 2. Depending on the load's demand, the DSTATCOM works in one of these modes. Assume the microgrid at steady-state condition and the DSTATCOM operating in mode-1. In this mode, the DSTATCOM supplies the total demand of the nonlinear load. At $t = 0.3$ s, the load demand increases and becomes larger than the capacity of renewable source of DSTATCOM. Therefore, the DSTATCOM starts to operate in mode-2 and it supplies half of the power demand of the load (assuming $\lambda_P = \lambda_Q = 0.5$) while the other half is injected by the microgrid feeder. At $t = 1.3$ s, the load demand reduces to the initial value and the DSTATCOM starts to operate in mode-1 again. Fig. (**4.38**) shows the injected active power by the DSTATCOM, the active power flow in the microgrid feeder, and the active power drawn by the load [14].

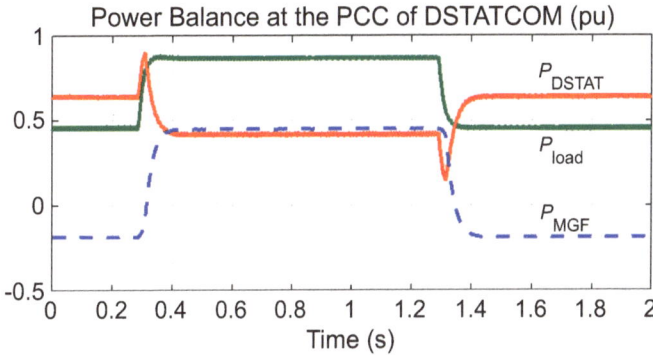

Fig. (4.38). Sample results for the system of Fig. (**4.27**), in the presence of a DSTATCOM operating in load compensation and energy harvesting mode.

Interphase Power Circulation in a Microgrid with an Unbalanced Generation

DSTATCOM can be used in microgrids for power circulation from one phase to another. This is a critical application for DSTATCOM in microgrids that have single-phase DERs, especially when they are distributed randomly and unequally among the phases of a three-phase system. On the other hand, in a low voltage microgrid, the loads are also single-phase and distributed among the phases unequally. Thereby, there should be a mechanism to circulate the excessive generation from the single-phase DERs of one phase to the phases with larger demands. DSTATCOM can help to realize this.

Fig. (**4.39**) shows a three-phase DSTATCOM which regulates its PCC to a three-phase balanced voltage of 1 pu [15]. Denoting the instantaneous power output of the feeder and DSTATCOM respectively as p_{MGF} and p_{DSTAT} and the instantaneous power flowing in the feeder at the downstream of the DSTATCOM as p, the instantaneous power balance equations at the PCC of the DSTATCOM can be expressed as

$$p_{\text{DSTAT},a} + p_{\text{MGF},a} - p_a = 0$$
$$p_{\text{DSTAT},b} + p_{\text{MGF},b} - p_b = 0 \qquad\qquad \textbf{(4.73)}$$
$$p_{\text{DSTAT},c} + p_{\text{MGF},c} - p_c = 0$$

Assuming that the upstream side of the microgrid feeder is disconnected, the powers in DSTATCOM upstream will be zero ($p_{\text{MGF},a} = p_{\text{MGF},b} = p_{\text{MGF},c} = 0$). Then, the powers at the PCC of the DSTATCOM are as $p_{\text{DSTAT},a} = p_a$, $p_{\text{DSTAT},b} = p_b$, and $p_{\text{DSTAT},c} = p_c$. On the other hand, since the DSTATCOM constitutes a closed boundary circuit, the sum of the powers exiting this closed boundary is zero (*i.e.*, $p_{\text{DSTAT},a} + p_{\text{DSTAT},b} + p_{\text{DSTAT},c} = 0$). Therefore, $p_a = -(p_b + p_c)$. From Fig. (**4.39**), it can be seen that $p_a = -p_{\text{DER},a}$, p_{DER} being the power generated by the single-phase DER. Hence, it can be seen that $p_{\text{DER},a} = p_b + p_c$. This shows that the loads in phase-b and c can be supplied by the single-phase DER of phase-a, which confirms the interphase power circulation through the DSTATCOM.

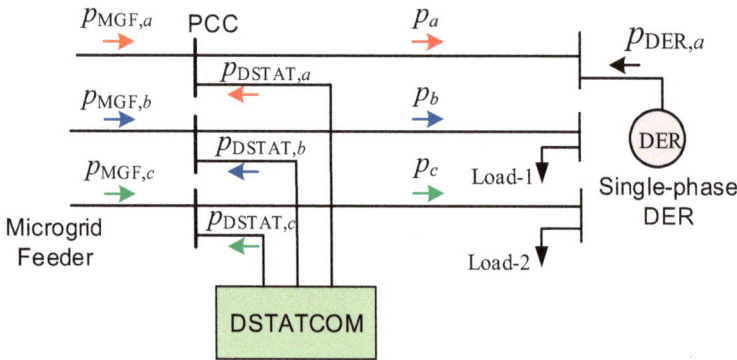

Fig. (4.39). Application of DSTATCOM for interphase power circulation in microgrids with single-phase DERs with unequal distribution amongst phases.

A current-controlled DSTATCOM with an LCL filter or a voltage-controlled DSTATCOM with an LC filter can be used for this purpose. The studies in [15, 31, 32] have shown that the current-controlled DSTATCOM is more suitable for medium voltage systems while a voltage-controlled DSTATCOM is more appropriate for low voltage systems. Other applications of DSTATCOM (such as voltage regulation or load compensation) can also be merged with this application to further the advantages of installing the DSTATCOM in the microgrid. The main important point is that the DSTATCOM should be composed of a common dc bus [15]. Thereby, the configurations of (Fig. **4.11**-**4.16**) are suitable for this purpose while the configurations of Figs. (**4.17** or **4.40**) are not appropriate for this application.

To principle of interphase power circulation by a DSTATCOM is the same as realizing a balanced power (and current) flow in the microgrid feeder. Thus, for a voltage-controlled DSTATCOM, the DSTATCOM has to regulate its PCC voltage to a set of balanced voltages. Thereby, for this DSTATCOM, the reference output voltages will be as given in (67) while for a current-controlled DSTATCOM, the references of the output current of this DSTATCOM will be as in (65).

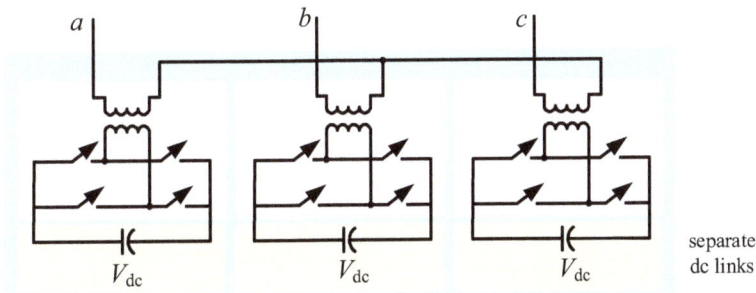

Fig. (4.40). A three-phase DSTATCOM with separate dc links that is unsuitable for interphase power circulation.

As an example, consider the system of Fig. (**4.27**) in which several single-phase DERs and loads are considered. A current-controlled DSTATCOM with an LCL filter (Fig. **4.21b**) is considered. Assume the system at steady-state condition when the DSTATCOM connects at $t = 0.5$ s. Fig. (**4.41**) illustrates the active power drawn from the microgrid feeder for each phase separately. As seen from this figure, phase-a and c observe a reverse active power flow while phase-b depicts a positive one. Thus, it can be concluded that the power generation by the single-phase DERs of phase-a and c is larger than the demand of these two phases while the demand of phase-b is larger than the generation of power by its single-phase DERs. Before the DSTATCOM connection, the first existing transformer in the upstream of the microgrid feeder would provide a power circulation if it has a delta connection or is constructed on a three-limb or 5-limb core [15].

The DSTATCOM forces the power drawn from the microgrid feeder to be balanced while the unbalanced phase power flow is limited to the downstream of the DSTATCOM. All phase powers become almost equal after the DSTATCOM connection as it can successfully circulate the powers among the phases. This is illustrated in Fig. (**4.41**) which shows that that one phase of the DSTATCOM absorbs power and returns it into the other two phases. It is to be noted that the algebraic sum of these powers will be equal to the power loss of the DSTATCOM. The current drawn from the microgrid feeder is also shown in this figure which illustrates the flow of large unbalanced currents before the DSTATCOM connection (because of a large amount of reverse power flow) while the magnitude of the currents reduces significantly after the DSTATCOM connection (as the microgrid feeder only injects the difference between the total generated power by single-phase DERs and the total consumption by all loads) [26].

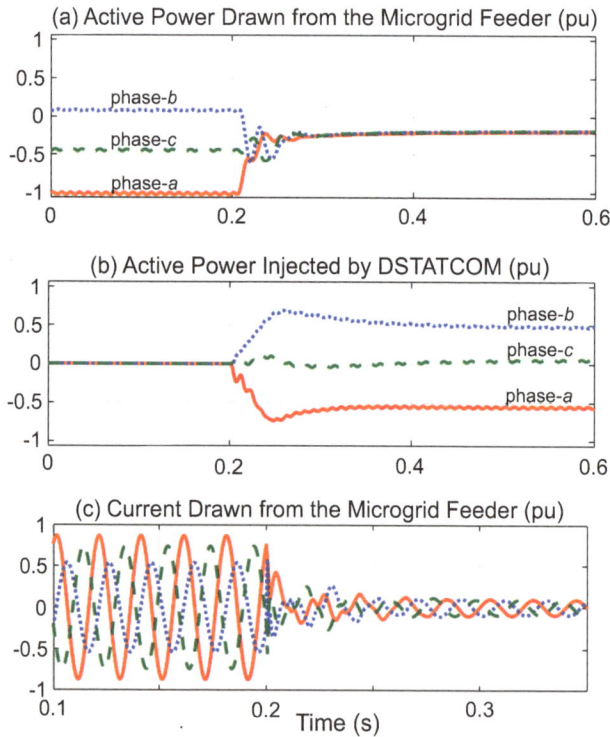

Fig. (4.41). Sample results for the system of (Fig. **4.27**), before and after the connection of a DSTATCOM operating in interphase power circulation mode.

SUMMARY

This chapter presented an overview of the operation and control principles of microgrids, the structure and control perspectives of DSTATCOMs, and some applications of DSTATCOMs in microgrids. In summary, DSTATCOMs can be used for load compensation, voltage regulation, and interphase power circulation in microgrids. These applications can also be merged with energy harvesting from renewable sources, when the DSTATCOM's dc link is coupled with an energy source. In brief, a voltage source converter with a path for the flow of zero sequence component of the current is preferred as it can facilitate the independent control of each phase of the system. This can be realized in center-tapped 3-leg VSCs, 4-leg VSCs or a three-phase DSTATCOM composed of three parallel single-phase full-bridges. A voltage-controlled DSTATCOM is more appropriate when the DSTATCOM intends to provide voltage regulation and balancing functions. In this condition, an LC filter is better to be used at the output of its VSC, and the voltage across the capacitor to be regulated to the desired voltage. On the other hand, a current-controlled DSTATCOM is more suitable when the DSTATCOM aims to improve the power factor or compensate the load (*i.e.*, eliminate its harmonic, and negative and zero sequence current components). In such conditions, an LCL filter is better to be used at the output of the VSC and the output current of the DSTATCOM to be regulated to the desired current. Both DSTATCOM schemes can be used for interphase power circulation in microgrids while a voltage controlled one is preferred for low voltage microgrids and a current-controlled one for medium voltage ones.

REFERENCES

[1]   MHJ Bollen, and F Hassan, *Integration of Distributed Generation in the Power System* Wiley-IEEE Press, 2011.

[2]   B. Lasseter, "Microgrids [distributed power generation]", *Power Engineering Society Winter Meeting,* 2001

[3]   F. Shahnia, "Hybrid nanogrid systems for future small communities", In: *Sustainable Development for Energy Systems.,* B. Azzopardi, Ed., Springer, 2016.

[4]   J.M. Guerrero, "Hierarchical control of droop-controlled dc and ac microgrids — a general approach towards standardization", *Industrial Electronics, 2009 IECON '09 35th Annual Conference of IEEE,* 2009

[5]   M.C. Chandorkar, D.M. Divan, and R. Adapa, "Control of parallel connected inverters in standalone ac supply systems", *IEEE Trans. Ind. Appl.,* vol. 29, no. 1, pp. 136-143, 1993. [http://dx.doi.org/10.1109/28.195899]

[6]   R.R. Kolluri, I. Mareels, T. Alpcan, and M. Brazil, "Power sharing correction in angle droop controlled inverter interfaced microgrids", In: *2015 IEEE Power & Energy Society General Meeting,* , 2015.

[7]   "Optimal angle droop power sharing control for autonomous microgrid", In: *2015 IEEE Energy Conversion Congress and Exposition.* ECCE, 2015.

[8]   T.L. Vandoorn, J.D. De Kooning, B. Meersman, and L. Vandevelde, "Review of primary control strategies for islanded microgrids with power-electronic interfaces", *Renew. Sustain. Energy Rev.,* vol. 19, pp. 613-628, 2013. [http://dx.doi.org/10.1016/j.rser.2012.11.062]

[9]   HL Willis, *Power Distribution Planning Reference Book* CRC Press Press, 2004.

[10]   C. Sankaran, *Power Quality.* CRC Press, 2001. [http://dx.doi.org/10.1201/9781420041026]

[11]   A. Ghosh, and G. Ledwich, *Power Quality Enhancement using Custom Power Devices.* Kluwer Academic Publishers, 2002.
[http://dx.doi.org/10.1007/978-1-4615-1153-3]

[12]   F. Shahnia, A. Ghosh, G. Ledwich, and F. Zare, "Voltage unbalance improvement in low voltage residential feeders with rooftop PVs using custom power devices", *Int. J. Electr. Power Energy Syst.,* vol. 55, pp. 362-377, 2014.
[http://dx.doi.org/10.1016/j.ijepes.2013.09.018]

[13]   F. Shahnia, S. Rajakaruna, and A. Ghosh, *Static Compensators (STATCOMs) in Power Systems.* Springer, 2015.
[http://dx.doi.org/10.1007/978-981-287-281-4]

[14]   F. Shahnia, R. Majumder, A. Ghosh, G. Ledwich, and F. Zare, "Operation and control of a hybrid microgrid containing unbalanced and nonlinear loads", *Electr. Power Syst. Res.,* vol. 80, no. 8, pp. 954-965, 2010.
[http://dx.doi.org/10.1016/j.epsr.2010.01.005]

[15]   F. Shahnia, and R.P. Chandrasena, "A three-phase community microgrid comprised of single-phase energy resources with an uneven scattering amongst phases", *Int. J. Electr. Power Energy Syst.,* vol. 84, pp. 267-283, 2017.
[http://dx.doi.org/10.1016/j.ijepes.2016.06.010]

[16]   F. Shahnia, R.P. Chandrasena, S. Rajakaruna, and A. Ghosh, "Primary control level of parallel distributed energy resources converters in system of multiple interconnected autonomous microgrids within self-healing networks", *IET Gener. Transm. Distrib.,* vol. 8, no. 2, pp. 203-222, 2014.
[http://dx.doi.org/10.1049/iet-gtd.2013.0126]

[17]   "Secondary control in microgrids for dynamic power sharing and voltage/frequency adjustment", *Power Engineering Conference (AUPEC),* 2014
[http://dx.doi.org/10.1109/AUPEC.2014.6966619]

[18]   E Pashajavid, F Shahnia, and A. Ghosh, "Development of a Self-Healing Strategy to Enhance the Overloading Resilience of Islanded Microgrids", *IEEE Transactions on Smart Grid,* vol. 8, no. 2, pp. 868-880, 2017.
[http://dx.doi.org/10.1109/TSG.2015.2477601]

[19]   F Shahnia, S Bourbour, and A. Ghosh, "Coupling Neighboring Microgrids for Overload Management Based on Dynamic Multicriteria Decision-Making", *IEEE Transactions on Smart Grid,* vol. 8, no. 2, pp. 969-983, 2017.
[http://dx.doi.org/10.1109/TSG.2015.2477845]

[20]   F. Shahnia, "Interconnected microgrid systems for remote areas", In: *Sustainable Development for Energy Systems.,* B. Azzopardi, Ed., Springer, 2016.

[21]   A Yazdani, and R. Iravani, *Voltage-Sourced Converters in Power Systems: Modeling, Control, and Applications,* Wiley-IEEE Press, 2010.

[22]   J.M. Guerrero, N. Berbel, J. Matas, and J.L. Sosa, "Droop Control Method with Virtual Output Impedance for Parallel Operation of Uninterruptible Power Supply Systems in a Microgrid", *APEC 07 - Twenty-Second Annual IEEE Applied Power Electronics Conference and Exposition,* 2007

[23]   "Different power sharing techniques for converter-interfaced DERs in an autonomous microgrid", *2014 IEEE PES Asia-Pacific Power and Energy Engineering Conference (APPEEC),* 2014
[http://dx.doi.org/10.1109/APPEEC.2014.7066196]

[24]   T. Hosseinimehr, F. Shahnia, A. Ghosh, Ed., *Dynamic power sharing control among converter-interfaced DERs in an autonomous microgrid.,* 2015. PowerTech, 2015 IEEE Eindhoven
[http://dx.doi.org/10.1109/PTC.2015.7232566]

[25]   "Power sharing control of batteries within autonomous microgrids based on their state of charge", *Power Engineering Conference (AUPEC),* 2015pp. 27-30

[http://dx.doi.org/10.1109/AUPEC.2015.7324848]

[26]   A. Ghosh, and F. Shahnia, "Applications of power electronic devices in distribution systems", In: *Transient Analysis of Power Systems: Solution Techniques, Tools and Applications: Wiley-IEEE Press.,* J.A. Martinez-Velasco, Ed., , 2015, pp. 248-279.
[http://dx.doi.org/10.1002/9781118694190.ch7]

[27]   E. Kabalci, "Converter and Output Filter Topologies for STATCOMs", In: *Static Compensators (STATCOMs) in Power Systems.,* F. Shahnia, S. Rajakaruna, A. Ghosh, Eds., Springer, 2015.
[http://dx.doi.org/10.1007/978-981-287-281-4_1]

[28]   M. Gopal, *Digital Control and State Variable Methods.* McGraw-Hill: Tata, 2009.

[29]   A. Tewari, *Modern Control Design with Matlab and Simulink.* Wiley, 2002.

[30]   F. Shahnia, R.P. Chandrasena, A. Ghosh, and S. Rajakaruna, "Application of DSTATCOM for surplus power circulation in MV and LV distribution networks with single-phase distributed energy resources", *2014 IEEE PES General Meeting Conference & Exposition,* 2014

[31]   F. Shahnia, S.M. Ami, and A. Ghosh, "Circulating the reverse flowing surplus power generated by single-phase DERs among the three phases of the distribution lines", *Int. J. Electr. Power Energy Syst.,* vol. 76, pp. 90-106, 2016.
[http://dx.doi.org/10.1016/j.ijepes.2015.09.021]

[32]   F. Shahnia, R.P. Chandrasena, A. Ghosh, and S. Rajakaruna, "Application of DSTATCOM for surplus power circulation in MV and LV distribution networks with single-phase distributed energy resources", *Electr. Power Syst. Res.,* vol. 117, pp. 104-114, 2014.
[http://dx.doi.org/10.1016/j.epsr.2014.08.010]

SUBJECT INDEX

A

AC network 5, 6, 7, 18, 22, 23, 34, 35, 37, 70
Active power control 14
Active power output 93
Adaptive neuro-fuzzy controller 21
Analogue-digital converter (ADC) 115, 116
Application of DSTATCOM 136
Application of SMES 59, 62, 64, 70
Auxiliary capacitors 106, 107

B

Battery energy storage (BES) 88, 91, 92, 99, 100, 101, 102
BES systems 99, 100, 101, 102

C

Capacitor voltage 115
Capacity 2, 3, 102, 108
 highest installed wind power 2
 increased power 108
 installed 2, 3
 installed wind power 2, 3
 maximum power 102
Capacity power converter 4
Compensators, static 17, 87, 89
Connection 23, 122
 single-phase/three-phase 122
 unity power factor 23
Control algorithm 53, 54
Control circuit 34, 35
Controller, primary 90, 91
Control parameters 53, 120
Control system, closed-loop 116, 121
Control techniques 88, 90, 102, 113
Converter, electronics-based 89
Converter faults 34, 70, 75
 internal 34
Converter faults event 50
Converter power 35
Converter station faults 43
Converter stations 7, 8, 22, 34, 50, 70, 75, 83

Converter switches 34, 105
Converter terminals 37, 38
Converter variable 6, 7
Coupling, common 8, 13, 17, 22, 23, 25, 49, 92
Coupling transformer 7, 22
Current-controlled DSTATCOM 113, 114, 116, 121, 122, 125, 131, 136, 137, 139
Current-voltage phase difference 133
Cut-off frequency 109, 110, 120

D

Damages, power grid 27
Damping 21, 28, 29, 42
Dc-dc chopper 52, 53, 55
Dc link 7, 50, 103, 104, 105, 108, 109, 110, 118, 121, 122, 123, 124, 126, 131, 137
 isolated 108
Dc link capacitors 22, 56, 106, 122
Dc link of DFIG 61, 63
Dc link voltage 36, 39, 70, 103, 121, 122, 123, 126
Dc link voltage of DFIG 36, 39
Dc load flow analysis 93
Dc voltage experiences 38
DERs, renewable energy-based 89, 91
DERs control 88
Deviation 29, 30, 31, 89, 122
 speed 29, 30, 31
Devices 11, 14, 17, 22, 34, 50, 75, 89
 custom power 89
DFIG and SMES coil 53
DFIG-based WECS 17, 22, 34, 42, 59, 62, 70, 83
DFIG dc-link 72, 74
DFIG dc link capacitor 61, 64
DFIG dc-link voltage 70

www.ingramcontent.com/pod-product-compliance
Lightning Source LLC
Chambersburg PA
CBHW041728210326
41598CB00008B/817

* 9 7 8 1 6 8 1 0 8 5 4 3 2 *